Economy of Life

Economy of Life

Linking Poverty, Wealth and Ecology

Edited by
Rogate R. Mshana and Athena Peralta

ECONOMY OF LIFE:
LINKING POVERTY, WEALTH AND ECOLOGY
Edited by Rogate R. Mshana and Athena Peralta

Trade edition 2015. Copyright © 2013 WCC Publications. All rights reserved. Except for brief quotations in notices or reviews, no part of this book may be reproduced in any manner without prior written permission from the publisher. Write: publications@wcc-coe.org.

WCC Publications is the book publishing programme of the World Council of Churches. Founded in 1948, the WCC promotes Christian unity in faith, witness and service for a just and peaceful world. A global fellowship, the WCC brings together more than 345 Protestant, Orthodox, Anglican and other churches representing more than 550 million Christians in 110 countries and works cooperatively with the Roman Catholic Church.

Opinions expressed in WCC Publications are those of the authors.

Cover design: Linda Hanna
Cover image: Paul Jeffrey
ISBN: 978-2-8254-1594-8

World Council of Churches
150 route de Ferney, P.O. Box 2100
1211 Geneva 2, Switzerland
http://publications.oikoumene.org

Contents

Preface vii

Economy of Life 1

Chapter 1 - *AGAPE: Background, Mandate and Context* 11

Chapter 2 - *Globalization: Political, Economic and Social Analysis* 15

Chapter 3 - *Issues of Faith: Theological Reflection* 35

Chapter 4 - *Dialogue on Economic Globalization: Religious, Economic and Political Actors* 43

Chapter 5 - *Sharing Practical Approaches: Areas of Concern* 57

Chapter 6 - *God of Life, Lead Us to Justice and Peace: A Conclusion* 69

Bibliography 71

Appendices

Appendix 1: PWE Consultations 75

Appendix 2: Statement on Eco-Justice and Ecological Debt 79

Appenidix 3: Statement on Just Finance and the Economy of Life 87

Appenidx 4: Statement on the Current Financial and Economic Crisis with a Focus on Greece 95

Appenix 5: The São Paulo Statement... 99

Abbreviations.. 116

Preface

This book is a harvest of the ecumenical work done to advance economic, social and ecological justice since the 9th Assembly of the World Council of Churches (WCC) in Porto Alegre till the 10th WCC Assembly in Busan. It outlines the milestones in the journey leading up to the Bogor Statement on "Economy of Life, Justice and Peace for All," also included in this volume. Let me outline what our mission was all about as part of the Alternative Globalization Addressing People and Earth (AGAPE) process focusing on poverty, wealth and ecology and their linkages.

The mission of the ecumenical movement today is about transforming the world into a place of justice and peace for all of God's creation. This understanding of mission has led to the theme of the 10th WCC Assembly, "God of Life, Lead us to Justice and Peace." The AGAPE process, which was an attempt by the ecumenical movement to define this way of mission, triggered a challenging debate among churches on what it means to be church today. On the other hand, it is also a debate about encouraging churches to put this issue at the centre of their mission and to point to alternatives to the current globalization of death.

The AGAPE process was launched at the 8th WCC Assembly in Harare, in 1998, with the question: "How do we live our faith in the context of globalization?" This question, which is central to Christian faith, will continue to be raised as long as a major part of humanity and the Earth suffer. The WCC's motto of preferential option for people in poverty remains relevant today. The initial phase of economic globalization was a period of euphoria with an assumption that salvation had come to humankind – a corporate globalized market was expected to eradicate poverty and bring prosperity for all. From the very beginning of this process, the WCC had made a clear distinction between globalization as a multi-faceted historic process and the present form of a pernicious economic and political project of global capitalism. The latter form of globalization is based on an ideology that groups and movements involved in the World Social Forum have described as "neoliberalism." The distinction between the two understandings of globalization was introduced by the Copenhagen Seminars for Social Progress.[1] The 8th WCC Assembly gave the mandate for a study on economic globalization. The results of this study process were disturbing.

At the 9th WCC Assembly in Porto Alegre, in 2006, an AGAPE Call was presented calling churches to action, urging them to work against economic injustice inherent in globalization. This call was issued at a time when the unfettered market took on virtually the image of god and when the role of governments to enforce regulations to protect public goods was severely weakened. Commerce and finance began to rule over politics, and governments were driven by corporate interests instead of peoples' needs. Financial markets were deregulated. Poor people and their nations were mercilessly left to the vagaries of the market. "Do not worry, the free market will bring prosperity for all" was the common mantra and remains so till today. Yet in 2008, we experienced the incredibly destructive impacts of neoliberal policies during the global financial crisis, from which the world has not recovered yet. The globalization of poverty and inequality has increased threats to peace.

Questioning globalization and seeking alternatives with the churches playing a central role is what the AGAPE process is about. This process is based on the understanding that reflection and action must continue by faithfully linking poverty, wealth and ecology as a concrete way of addressing economic, social and ecological injustice. Together with the covenanting for justice process of the World Alliance of Reformed Churches, now called World Communion of Reformed Churches (WCRC), the AGAPE process has given birth to the Oikotree movement which is encouraging ecumenical movements to focus on putting justice at the heart of faith.

It is evident that individual churches cannot effectively work against economic injustice in the current frame of globalization. They need to work ecumenically and with social movements as a new way of contributing to a fresh understanding of ecumenism today.

---

[1] Jacques Baudot, ed. (2000), *Building a World Community. Globalization and the Common Good*, Royal Danish Ministry of Foreign Affairs: Copenhagen.

One of the main concerns about globalization is almost certainly the public feeling, supported by empirical evidence, that the gaps between rich and the poor among and within countries are widening. Yet another concern is the merciless mutilation of the Earth in the search of profits such that the productive and regenerative limits of the planet have now been overstepped. This scenario will not change if churches stop raising their voices against the dominant capitalist economic model. Church campaigns on debt cancellation for poor countries, fair trade, climate change mitigation and adaptation, making poverty history, and promoting the right to food are signs of hope. But these should not bring contentment to church workers and the ecumenical movement. Often such an approach assumes that the problems globalization generates are problems for the South. This is not true – the 2008 global financial and economic crash and the reality of climate change highlight our common vulnerability as a human and ecological community. Churches should not confine themselves to designing projects that will help poor people in the South. Nor should they believe that capitalism as an economic system is beyond interrogation.

Regrettably, some churches have fallen in line with corporate thinking and are also using the language of money and markets. While some might be excited to speak for instance about aid effectiveness, others should raise the question as to why there is a need for aid in the first place. A rice farmer whose livelihood is destroyed because of the influx of imported and subsidized rice from the U.S. or a public health facility that has scaled down as a result of government budget cuts may need aid in the short run to survive. But the fundamental issues to be tackled are injustices in global trade and financial systems including relentless capital flight from poor to rich countries.

What is needed is for churches to work for global justice by critically and prophetically addressing capitalism, which, despite numerous mutations, cannot eradicate poverty, inequality and ecological destruction. This conclusion was reached by many of the study processes and consultations on poverty, wealth and ecology involving more than 500 participants, including church representatives, women, youth, Indigenous People and theologians, organized from 2007 to 2013 by the WCC Poverty, Wealth and Ecology (PWE) project. The answer lies in a participatory search for alternatives that are centred on the people and the Earth.

The various activities under the PWE project on critiquing the role of the European Union, United States and China in Africa, on recognizing ecological debt and developing forms of reparation, on elaborating a greed line and multidimensional indicators of greed, on transforming the international financial architecture, etc., have been attempts to bring to the fore areas that require further work and follow ups. Since the 9th WCC Assembly in Porto Alegre, we have come far as churches in terms of arriving at a more common understanding and shared stance on globalization, especially as some of us were severely affected by the 2008 global financial crisis. The statement on "Economy of Life, Justice and Peace for All"[2] coming from the Global Forum on PWE held in Bogor, Indonesia in 2012, summarizes this journey and is a contribution to the Pilgrimage for Justice and Peace, which is expected to be launched after the 10th Assembly in Busan.

It is our hope that this document will serve as a landmark of where churches stand in the work on promoting an Economy of Life and in the search for alternatives to the current paradigm of capitalism. Continuing inequities in the global distribution of resources and in the workings of trade and financial systems as well as unlimited and unsustainable production and consumption are the root-causes of the intertwined socioeconomic and ecological crises we face today. This great agenda must be at the ecumenical table as part of the mission for public witness and service of the WCC. The WCC must consistently raise issues of global justice because no one else will do so at the prophetic level and from the understanding that these lie at the very heart of our faith.

---

[2] The statements mentioned in the preface and the acknowledgements can all be found in the bibliography of this report. The statements "Just Finance and the Economy of Life," "Eco-Justice and Ecological Debt.," and "International Financial Transformation for an Economy of Life" are also included as appendices in this volume.

Acknowledgments

First and foremost, we express our gratitude to partners who resourced the PWE project in a large scale, namely: the United Church of Canada, the Canadian International Development Agency, the Mission Covenant Church of Sweden, World Mission and the Mennonite Church in the Netherlands. A special thanks to ICCO and Bread for the World, who enabled us to convene the WCC Advisory Group on Economic Matters (AGEM) to discuss the financial crisis, resulting in the publication entitled Justice Not Greed.

The work in this field was staffed by one person based at the WCC headquarters in Geneva with 35 percent administrative support and a consultant based in Manila. A big 'thank you' must be extended to Sophie Dhanjal and Alexandra Pomezny for their indefatigable assistance and efficient administration of the project. Moreover, it would have been difficult if not impossible to have accomplished the same quantity and quality of work if not for our consultant working from the Philippines, Athena Peralta, to whom we register our appreciation.

We are grateful to the AGAPE-PWE reference group composed of representatives of regional ecumenical organizations, churches and development agencies that helped to implement the programme effectively. In particular we would like to thank the All Africa Conference of Churches (Arthur Shoo), Council of Latin American Churches (Franklin Canelos), Christian Conference of Asia (Charlie Ocampo), and the Conference of European Churches (Peter Pavlovic) that hosted and organized our meetings and consultations.

We are also indebted to our researchers from Africa (Clement Kwayu), Latin America and the Caribbean (Carlos Larrea, Jorge Atillio Silva Iulianelli, Jorge Coronado Marroquin), Asia and the Pacific (Rosario Bella Guzman), Europe (coordinated by Peter Pavlovic) and North America (Elizabeth Hinson-Hasty and John Dillon). We have not forgotten Dr Aruna Gnanadason who provided advice and raised the issue of women, economy and ecology.

We further acknowledge the incredibly enriching work of the various study teams that supported the PWE process, namely: the AGEM (led by Jan Pronk), the Ecological Debt Group (especially Joy Kennedy, William Stanley, Malcolm Damon and Aurora Donoso), and the Greed Line Study Group (led by Konrad Raiser, former WCC General Secretary).These working groups were central to the formulation of the WCC central committee statements on "Just Finance and the Economy of Life" and "Eco-Justice and Ecological Debt." We are also grateful to the WCC central committee which issued these statements.

Many thanks to the WCC publications team, the Director of Communications, the website editor and all colleagues at the WCC who gave support to this work.

Finally, we would like to show our appreciation to the WCRC, the Council for World Mission (CWM) and the Lutheran World Federation (LWF). We partnered with the LWF to discuss interfaith perspectives on how to overcome greed. Together with the WCRC and the CWM, we initiated the Oikotree movement and came up with the São Paulo statement on "International Financial Transformation for an Economy of Life." Working ecumenically has been inspiring and rewarding. May this way of working be sustained.

In summary, many people and institutions offered a hand in bringing the WCC's critical work on advancing justice in the economy and the Earth to a level that is satisfying. We are most thankful for their contributions.

Dr Rogate R. Mshana

Economy of Life

As a follow-up to the Alternative Globalization Addressing People and Earth (AGAPE) process, which concluded with the AGAPE Call presented at the 9th Assembly of the World Council of Churches (WCC) in Porto Alegre in 2006, the WCC initiated a programme focused on eradicating poverty, challenging wealth accumulation and safeguarding ecological integrity based on the understanding that Poverty, Wealth and Ecology (PWE) are integrally related. The PWE programme engaged in on-going dialogue between religious, economic and political actors. Participants included ecumenical leaders, representatives and leaders of churches from all over the world, interfaith partners, leaders of government and social service organizations, and it represented a rich variety of the world's regions and nations. Regional studies and consultations took place in Africa (Dar es Salaam) in 2007, Latin America and the Caribbean (Guatemala City) in 2008, Asia and the Pacific (Chiang Mai) in 2009, Europe (Budapest) in 2010 and North America (Calgary) in 2011. The programme culminated in a Global Forum and AGAPE celebration in Bogor, Indonesia in 2012. The call to action that follows is the result of a six-year process of consultations and regional studies linking poverty, wealth and ecology.

Preamble

1. This call to action comes in a time of dire necessity. People and the Earth are in peril due to the over-consumption of some, growing inequalities as evidenced in the persistent poverty of many in contrast to the extravagant wealth of a few, and intertwined global financial, socio-economic, ecological and climate crises. Throughout our dialogue, we as participants in consultations and regional studies expressed differing, sometimes even contrasting, perspectives. We also grew to share a common consciousness that life in the global community as we know it today will come to an end if we fail to confront the sins of egotism, callous disregard and greed which lie at the root of these crises. With a sense of urgency, we bring this dialogue to the churches as a call to action. This urgency is born of our profound hope and belief: An Economy of Life is not only possible, it is in the making—and God's justice lies at its very foundation!

Theological and Spiritual Affirmations of Life

2. The belief that God created human beings as part of a larger web of life and affirmed the goodness of the whole creation (Genesis 1) lies at the heart of biblical faith. The whole community of living organisms that grows and flourishes is an expression of God's will and works together to bring life from and give life to the land, to connect one generation to the next, and to sustain the abundance and diversity of God's household (*oikos*). Economy in God's household emerges from God's gracious offering of abundant life for *all* (John 10:10). We are inspired by Indigenous Peoples' image of "Land is Life" (*Macliing Dulag*) which recognizes that the lives of people and the land are woven together in mutual interdependence. Thus, we express our belief that the "creation's life and God's life are intertwined" (Commission on World Mission and Evangelism) and that God will be all in all (1 Corinthians 15:28).

3. Christian and many other expressions of spirituality teach us that the "good life" lies not in the competitive quest for possessions, the accumulation of wealth, fortresses and stockpiles of armaments to provide for our security, or in using our own power to lord it over others (James 3: 13-18). We affirm the "good life" (*Sumak Kausay* in the Kichua language and the concept of *Waniambi a Tobati Engros* from West Papua) modeled by the communion of the Trinity in mutuality, shared partnership, reciprocity, justice and loving-kindness.

4. The groaning of the Creation and the cries of people in poverty (Jeremiah 14:2-7) alert us to just how much our current social, political, economic and ecological state of emergency runs counter to God's vision for life in abundance. Many of us too easily deceive ourselves into thinking that human desires stand at the centre of God's universe. We construct divisions, barriers and boundaries to distance ourselves from neighbour, nature and God's justice. Communities are fragmented and relationships broken. Our greed and self-centredness endanger both people and planet Earth.

5. We are called to turn away from works that bring death and to be transformed into a new life (*metanoia*). Jesus calls humanity to repent of our sins of greed and egotism, to renew our relationships with the others and creation, to restore the image of God, and to begin a new way of life as a partner of God's life-affirming mission. The call of the prophets is heard anew from and through people submerged in poverty by our current economic system and those most affected by climate change: Do justice and bring a new Earth into being!

6. Our vision of justice is rooted in God's self-revelation in Jesus Christ who drove money changers from the temple (Matthew 21:12), made the weak strong and strong weak (1 Corinthians 1:25-28), and redefined views of poverty and wealth (2 Corinthians 8:9). Jesus identified himself with the marginalized and excluded people not only out of compassion, but because their lives testified to the sinfulness of the systems and structures. Our faith compels us to seek justice, to witness to the presence of God and to be part of the lives and struggles of people made weak and vulnerable by structures and cultures—women, children, people living in poverty in both urban and rural areas, Indigenous Peoples, racially oppressed communities, people with disabilities, Dalits, forced migrant workers, refugees and religious ethnic minorities. Jesus says "Whatever you did to the least of these you did to me" (Matthew 25: 40).

7. We must embody a "transformative spirituality" (Commission on World Mission and Evangelism) that re-connects us to others (*Ubuntu* and *Sansaeng*), motivates us to serve the common good, emboldens us to stand against all forms of marginalization, seeks the redemption of the whole Earth, resists life-destroying values and inspires us to discover innovative alternatives. This spirituality provides the means to discover the grace to be satisfied with enough, while sharing with any who have need (Acts 4: 35).

8. Churches must be challenged to remember, hear and heed Christ's call today: "The time has come … The kingdom of God is near. Repent and believe the good news!" (Mark 1:15). We are called to

be transformed, to continue Christ's acts of healing and reconciliation and "to be what [we] have been sent to be—a people of God and a community in the world" (Poverty, Wealth, and Ecology in Africa). Therefore, the Church is God's agent for transformation. The Church is a community of disciples of Jesus Christ, who affirms the fullness of life for all, against any denial of life.

Intertwined and Urgent Crises

9. Our present stark global reality is so fraught with death and destruction that we will not have a future to speak of unless the prevailing development paradigm is radically transformed and justice and sustainability become the driving force for the economy, society and the Earth. Time is running out.

10. We discern the fatal intertwining of the global financial, socio-economic, climate, and ecological crises accompanied in many places of the world by the suffering of people and their struggle for life. Far-reaching market liberalization, deregulation and unrestrained privatization of goods and services are exploiting the whole Creation and dismantling social programmes and services and opening up economies across borders to seemingly limitless growth of production. Uncontrolled financial flows destabilize the economies of an increasing number of countries all over the world. Various aspects of climate, ecological, financial and debt crises are mutually dependent and reinforce each other. They cannot be treated separately anymore.

11. Climate change and threats to the integrity of creation have become the significant challenge of the multifaceted crises that we have to confront. Climate change directly impacts peoples' livelihoods, endangers the existence of small island states, reduces the availability of fresh water and diminishes Earth's biodiversity. It has far-reaching impacts on food security, the health of people and the living habits of growing part of population. Due to climate change, life in its many forms as we know it can be irreversibly changed within the span of a few decades. Climate change leads to the displacement of people, to the increase of forced climate

migration, and to armed conflicts. Unprecedented challenges of climate change go hand-in-hand with the uncontrolled exploitation of natural resources and leads to the destruction of the Earth and to a substantial change of the habitat. Global warming and ecological destruction become more and more a question of life or death.

12. Our world has never been more prosperous, and, at the same time, more inequitable than it is today. Inequality has reached a level that we can no longer afford to ignore. People who have been submerged into poverty, driven into overwhelming debt, marginalized, and displaced are crying out with a greater sense of urgency and clarity than ever before. The global community must recognize the need for all of us to join hands together and to do justice in the face of unparalleled and catastrophic inequalities in the distribution of wealth.

13. Greed and injustice, seeking easy profit, unjust privileges and short-term advantages at the expense of long term and sustainable aims are root causes of the intertwined crises and cannot be overlooked. These life-destroying values have slowly crept in to dominate today's structures and lead to lifestyles that fundamentally defy the regenerative limits of the Earth and the rights of human beings and other forms of life. Therefore, the crisis has deep moral and existential dimensions. The challenges that are posed are not first and foremost technological and financial, but ethical and spiritual.

14. Market fundamentalism is more than an economic paradigm: it is a social and moral philosophy. During the last thirty years, market faith based on unbridled competition and expressed by calculating and monetizing all aspects of life has overwhelmed and determined the direction of our systems of knowledge, science, technology, public opinion, media and even education. This dominating approach has funneled wealth primarily toward those who are already rich and allowed humans to plunder resources of the natural world far beyond its limits to increase their own wealth. The neoliberal paradigm lacks the self-regulating mechanisms to

deal with the chaos it creates with far-reaching impacts, especially for the impoverished and marginalized.

15. This ideology is permeating all features of life, destroying it from the inside as well as from the outside, as it seeps into the lives of families and local communities, wreaks havoc upon the natural environment and traditional life-forms and cultures, and spoils the future of the Earth. The dominant global economic system in this way threatens to put an end to both the conditions for peaceful coexistence and life as we know it.

16. The one-sided belief that social benefits automatically follow from economic (GDP) growth is misguided. Economic growth without constraints strangles the flourishing of our own natural habitat: climate change, deforestation, ocean acidification, biodiversity loss and so on. The ecological commons have been degraded and appropriated, through the use of military force, by the political and economic elite. Over-consumption based on the costs of uncovered debts generates massive social and ecological indebtedness, which are owed by the developed countries of global North to the global South, as well as indebtedness over against the Earth, is unjust and creates enormous pressure on future generations. The notion that the Earth is the Lord's and everything in it (Psalm 24: 1; 1 Corinthians 10: 26) has been dismissed.

Well-Springs of Justice

17. We confess that churches and church members are complicit in the unjust system when they partake in unsustainable lifestyles and patterns of consumption and remain entangled in the economy of greed. There are churches who continue to preach theologies of prosperity, self-righteousness, domination, individualism and convenience. Some support theologies of charity rather than justice for the impoverished. Others fail to question and even legitimize systems and ideologies founded on unlimited growth and accumulation, and ignore the reality of ecological destruction and the plights of victims of globalization. Some focus on short-term, quantifiable results at the expense of deep-seated, qualitative changes. However, we are also aware that even when many fail to

examine and change their own production, consumption and investment behaviour, an increasing number of churches from all continents are stepping up their efforts and expressing their belief that transformation is possible.

18. Ultimately, our hope springs from Christ's resurrection and promise of life for all. We see evidence of that resurrection hope in the churches and movements committed to making a better world. They are the light and salt of the Earth. We are profoundly inspired by numerous examples of transformation from within the family of churches and in growing movements of women, people in poverty, youth, people with disabilities and Indigenous Peoples who are building an Economy of Life and promoting a flourishing ecology.

19. People of faith, Christian, Muslim and Indigenous leaders in the Philippines, have given their lives to maintain their connection to and to continue to sustain themselves from the land to which they belong. Churches in South America, Africa and Asia are conducting audits of external debts and challenging mining and resource-extractive companies to be accountable for human rights violations and environmental damages. Churches in Latin America and Europe are sharing and learning from differing experiences with globalization and working toward defining common but differentiated responsibilities, building solidarity and strategic alliances. Christians are defining indicators of greed and conducting intentional dialogues with Buddhists and Muslims which discover common ground in the fight against greed. Churches in partnership with civil society are engaged in discussing the parameters of a new international financial and economic architecture, promoting life-giving agriculture and building economies of solidarity.

20. Women have been developing feminist theologies that challenge patriarchal systems of domination as well as feminist economics that embed the economy in society and society in ecology. Youth are in the forefront of campaigns for simple living and alternative lifestyles. Indigenous Peoples are making demands for holistic

reparations and the recognition of Earth rights to address social and ecological debt.

Commitments and Call

21. The 10thAssembly of the WCC is meeting at a time when the vibrant life of God's whole creation may be extinguished by human methods of wealth creation. God calls us to a radical transformation. Transformation will not be without sacrifice and risk, but our faith in Christ demands that we commit ourselves to be transformative churches and transformative congregations. We must cultivate the moral courage necessary to witness to a spirituality of justice and sustainability, and build a prophetic movement for an Economy of Life for all. This entails mobilizing people and communities, providing the required resources (funds, time and capacities), and developing more cohesive and coordinated programmes geared toward transforming economic systems, production, distribution, and consumption patterns, cultures and values.

22. The process of transformation must uphold human rights, human dignity and human accountability to all of God's creation. We have a responsibility that lies beyond our individual selves and national interests to create sustainable structures that will allow future generations to have enough. Transformation must embrace those who suffer the most from systemic marginalization, such as people in poverty, women, Indigenous Peoples and persons living with disabilities. Nothing determined without them is for them. We must challenge ourselves and overcome structures and cultures of domination and self-destruction that are rending the social and ecological fabric of life. Transformation must be guided by the mission to heal and renew the whole creation.

23. Therefore, we call on the 10th Assembly in Busan to commit to strengthening the role of the WCC in convening churches, building a common voice, fostering ecumenical cooperation and ensuring greater coherence for the realization of an Economy of Life for all. In particular, the critical work on building a new international financial and economic architecture (WCC Statement

on Just Finance and an Economy of Life), challenging wealth accumulation and systemic greed and promoting anti-greed measures (Report of the Greed Line Study Group), redressing ecological debt and advancing eco-justice (WCC Statement on Eco-justice and Ecological Debt) must be prioritized and further deepened in the coming years.

24. We further call on the 10th WCC Assembly in Busan to set aside a period of time between now and the next Assembly for churches to focus on faith commitments to an "Economy of Life – Living for God's Justice in Creation [Justice and Peace for All]." The process will enable the fellowship of churches to derive fortitude and hope from each other, strengthen unity and deepen common witness on critical issues that lie at the very core of our faith.

25. The statement on "Just Finance and an Economy of Life" calls for an ethical, just and democratic international financial regime "grounded on a framework of common values: honesty, social justice, human dignity, mutual accountability and ecological sustainability" (WCC Statement on Just Finance and an Economy of Life). We can and must shape an Economy of Life that engenders participation for all in decision-making processes that impact lives, provides for people's basic needs through just livelihoods, values and supports social reproduction and care work done primarily by women, and protects and preserves the air, water, land, and energy sources that are necessary to sustain life (Poverty, Wealth, and Ecology in Asia and the Pacific). The realization of an Economy of Life will entail a range of strategies and methodologies, including, but not limited to: critical self-reflection and radical spiritual renewal; rights-based approaches; the creation and multiplication of spaces for the voices of the marginalized to be heard in as many arenas as possible; open dialogue between global North and global South, between churches, civil society and state actors, and among various disciplines and world faiths to build synergies for resistance to structures and cultures that deny life in dignity for many; taxation justice; and the organization of a broad platform for common witness and advocacy.

26. The process is envisioned as a flourishing space where churches can learn from each other and from other faith traditions and social movements about how a transformative spirituality can counter and resist life-destroying values and overcome complicity in the economy of greed. It will be a space to learn what an Economy of Life means, theologically and practically, by reflecting together and sharing what concrete changes are needed in various contexts. It will be a space to develop joint campaigns and advocacy activities at the national, regional and global levels with a view to enabling policy and systemic changes leading to poverty eradication and wealth redistribution; ecologically-respectful production, consumption and distribution; and to develop healthy, equitable, post-fossil fuel and peace-loving societies.

God of Life calls us to justice and peace.
Come to God's table of sharing!
Come to God's table of life!
Come to God's table of love!

1.
AGAPE: Background, Mandate and Context

The AGAPE Process

The starting point is the question first raised at the 8th Assembly of the World Council of Churches (WCC) in Harare in 1998: How do we live our faith in the context of economic globalization? Building on the Alternative Globalization Addressing People and Earth (AGAPE) process, which was set in motion in Harare climaxing with the AGAPE Call presented at the 9th WCC Assembly in Porto Alegre in 2006, the ecumenical study and deliberation on the impacts of and responses to economic globalization were further deepened through the implementation of the Poverty, Wealth and Ecology (PWE) project from 2007 till 2012.

Understanding that poverty, wealth and ecology are integrally related, the PWE project addresses the need for economic transformation in the areas mentioned in the AGAPE Call:

- poverty eradication
- trade
- finance
- sustainable use of land and natural resources
- public goods and services
- life-giving agriculture
- decent jobs and people's livelihoods and
- the power of empire[1]

---
[1] For the complete document of the AGAPE Call see: www.oikoumene.org.

Lessons Learned:
Need for Open Dialogue and Critical Self-Reflection among Churches

Since the 1966 World Conference on Church and Society in Geneva, the economic sphere of life – particularly the ethical, moral and theological questions around control of the means of production, access to and distribution of resources and material goods – has been an important subject of rumination and contention for the churches.

The AGAPE process from 1998 to 2006 challenged churches around the world to interrogate the phenomenon of economic globalization and its impacts on people and on the Earth. Since then, the 2006 AGAPE Call and background document have been translated into at least 13 languages, discussed by churches and taught in some theological seminaries.

The AGAPE process, however, exposed continuing North–South divergences in analyses and recommendations among churches and ecumenical partners, stemming largely from contrasting worldviews and experiences with economic globalization which presents multifaceted, complex realities. The process also enabled the realization that churches are themselves complicit in dominant economic structures as producers, consumers and investors. The Accra Confession (WCRC, formerly WARC, 2004) states: "We acknowledge the complicity and guilt of those who consciously or unconsciously benefit from the current neoliberal economic global system; we recognize that this includes…churches…."

These challenges underscored the need for genuinely open and connected dialogue as well as critical self-reflection as churches on the phenomenon of economic globalization to strengthen understanding and synergies among churches and the ecumenical community so as to fulfill the threefold vision of the WCC of living out Christian unity more fully, being "neighbours to all" and taking great care of creation.

Mandate from the 9th WCC Assembly

The PWE project was born out of the 9th WCC Assembly in Porto Alegre, which approved a continuation of the AGAPE process in collaboration with other ecumenical partners and organizations with proposals to engage in:

- solid political, economic and social analysis;
- the work of theological reflection on these issues that arise out of the centre of our faith;
- ongoing dialogue between religious, economic and political actors; and
- sharing practical, positive approaches from the churches.

Comprehending the intrinsic connections between justice and peace, the WCC central committee, meeting in September 2007, further recommended a visible link to be established between the AGAPE-PWE process and the Decade to Overcome Violence and Just Peace Declaration.

Against this background, the objective of this report is to discuss how the PWE project has implemented the mandate from the Porto Alegre Assembly, highlighting the breakthroughs as well as the challenges faced in addressing the areas mentioned in the 2006 AGAPE Call.

The project spearheaded a series of regional researches and church consultations which took place in Africa (Dar es Salaam) in 2007, Latin America and the Caribbean (Guatemala City) in 2008, Asia and the Pacific (Chiang Mai) in 2009, Europe (Budapest) in 2010 and North America (Calgary) in 2011, culminating in the Global Forum on PWE held in Bogor in 2012 (see PWE Consultations, 38ff). It was guided by an advisory group comprised of representatives from churches and further supported by study teams namely: the Advisory Group on Economic Matters (AGEM) dealing with issues of global finance, the Ecological Debt Reference Group and the Greed Line Study Group.

2.

Globalization:
Political, Economic and Social Analysis

Connections between Poverty, Wealth and Ecology

Working on the recognition that poverty, wealth and ecology are intrinsically connected, the PWE project embarked on the conduct of research at various levels to better understand political, economic, and social trends in the current milieu characterized by globalization. Notably, the period 2007-2012 was one marked by unprecedented and intertwined global financial, economic and ecological emergencies, providing important openings for a sharper analysis of the interrelationships between poverty, wealth accumulation and ecological health. Specifically, there were attempts to find answers to the following sets of questions:

- To what extent are methods and structures of wealth creation responsible for poverty and inequality? What are concrete examples from countries, regions and worldwide that illustrate how the poor are deprived of their entitlements by the rich? How can this trend be reversed? How can wealth be shared equitably within countries and globally?

- To what extent does wealth creation generate environmental destruction? How do corporations in pursuit of profit generate ecological debt through their production methods? How do individuals contribute to ecological debt through their lifestyles and consumption patterns? How can ecological debt be redressed?

- What motivates human beings to accumulate more wealth than they really need? Should there be a limit beyond which the accumulation of wealth becomes greed and is no longer ethically and socially acceptable, i.e., a "greed line"? How realistic is it to talk about an "economy of enough"?

- Are inequality and the lack of wealth distribution threats to peace?

Some of the key findings arising from the study process as well as undertakings to act on these findings are shared below.

Poverty amidst Wealth: Growing Inequality and the Irony of Growth

We live in a world of paradoxes. On the one hand, the world has never been wealthier: in spite of the global financial and economic crisis of 2008, global wealth grew to US$120 trillion in 2010 or an increase of 20 percent from 2007 (Pushra and Burke 2011). Yet, on the other hand, the world has never been more unequal.

In "Poverty, Social Inequality and the Environment in Ecuador and Latin America,"[1] Carlos Larrea (in Mshana, ed. 2009) observes that "social inequality between countries in the world has grown, with an increase of the Gini coefficient from 0.47 in 1980 to 0.52 in 2000."[2]

> Inequality is higher for wealth distribution than for income distribution. More specifically, the richest one percent of the world's people holds 31 percent of the world's wealth while the richest 10 percent has 71 percent. At the opposite pole, the poorest half of the global population possesses merely 3.7 percent of global riches and the poorest ten percent has scarcely one thousandth of global wealth, translating to a lack of access to basic needs necessary for survival and a dignified life. As confirmed by some of the regional studies and consultations, the experience of impoverishment is further shaped by social hierarchies based on gender, class, race and ethnicity.[3]

---

[1] See *Poverty, Wealth and Ecology: Ecumenical Perspectives from Latin America and the Caribbean* edited by Rogate Mshana (2009).
[2] The Gini coefficient measures inequality (e.g., in levels of income). A coefficient of 0 expresses perfect equality (e.g., where everyone has an exactly equal income) while a coefficient of 1 expresses maximal inequality (e.g., where only one person has all the income).
[3] See Larrea, *Poverty, Social Inequality and the Environment in Ecuador and Latin America* (in Mshana, ed. 2009), 30.

The rise in inequality is linked to the growth paradox: even where wealth creation at the national level has occurred – as reflected in economic growth – this has not necessarily led to poverty reduction. Rosario Guzman (in Peralta, ed., 2010) in "Establishing the Links between Wealth Creation, Poverty, and Ecological Devastation: The Asian Experience"[4] points out that:

> Particularly in Asia, the emptiness of high rates of economic growth as measured by increases in gross domestic product (GDP) had long been betrayed by job insecurity, falling wages, disappearing social services, fluctuations in the prices of food and other basic commodities, and depleted natural riches. If the region was experiencing growth, it was exclusionary, deepening poverty for many…

How might one explain the sharp increase in inequality over a period of less than two generations? In "'As Any Might Have Need' – Envisioning Communities of Shared Partnership," Elizabeth Hinson-Hasty (in Kennedy, ed. 2012) writes:

> There is an emerging consensus…that neoliberalism, a particular form of capitalism that emerged in the 1970s, is the root cause of wealth inequalities and the patterns of consumption that are leading to the ecological destruction we are experiencing today. Neoliberalism has been the primary ideological fuel for changes in economic policies, the shaping of public perceptions and ideas, and the reshaping of social and political institutions…[5]

Indeed, over the last 30 years, the global shift in the political and economic agenda toward neoliberal policies promoting market liberalization, deregulation and privatization resulted in many instances

---

[4] See *Poverty, Wealth and Ecology in Asia and the Pacific: Ecumenical Perspectives* edited by Athena Peralta (2010)
(http://www.cca.org.hk/resource/books/olbooks/poverty_wealth_and_ecology_in_ap.pdf).
[5] The study is available at http://www.kairoscanada.org/wp-content/uploads/2011/11/SUS-CJ-11-10-Hinson-Hasty.pdf.

in the systematic transfer of wealth from the poor and middle-class majority to the already rich and powerful minority.

In *"Poverty, Wealth and Ecology in Brazil,"*[6] Jorge Atilio Silva Iullanielli (in Mshana, ed. 2009) uses the example of external and internal debt to demonstrate how wealth concentration globally and locally has been increasingly based on the privatization of social surpluses and the socialization of costs.

Notably, however, "debt is becoming a critical issue not only for poor and developing countries, but equally for wealthy countries" (Pavlovic, ed. 2011). The ongoing Eurozone sovereign debt crisis underlines this point.

In *"Poverty, Wealth and Ecological Debt in Central America,"*[7] Jorge Coronado Marroquin (in Mshana, ed. 2009) shows how agricultural trade liberalization in Central America led to the massive influx of food stuff from the United States, placed domestic agricultural markets under the control of transnational corporations, and displaced local farmers in Costa Rica, El Salvador, Guatemala, Nicaragua, and Panama, such that the Central American countryside has become "a reservoir of the marginalized."

Both Elizabeth Hinson-Hasty in her study on poverty, wealth and ecology in the United States and John Dillon (in Kennedy, ed. 2012) in "Poverty, Wealth and Ecology in Canada"[8] trace increasing socio-economic gaps in the United States and Canada in part to regressive taxation policies that benefited big businesses and affluent classes.

Reading the women's statements from the Africa, Latin America and Caribbean and Asia-Pacific consultations on PWE, it is evident that the impacts of neoliberal adjustment have been particularly harsh on women (see Box 1). And yet, at the same time, the Guatemala Declaration on

---

[6] See *Poverty, Wealth and Ecology: Ecumenical Perspectives from Latin America and the Caribbean* edited by Rogate Mshana (2009).
[7] Ibid.
[8] The study may be accessed at http://www.kairoscanada.org/wp-content/uploads/2011/11/SUS-CJ-11-10-PovertyWealthEcology.pdf.

PWE [9] observes that women's "invisible and unrecognized domestic work subsidizes the global economic model."

> 1. Impacts of neoliberal economic policies on women
>
> The interaction between patriarchy and neoliberal economic globalization has concentrated decision-making power and productive means and resources, especially capital, in the hands of the so-called "Davos man." It has also resulted in drastic cuts in investment in life-giving areas such as sustainable agriculture and education and health services. Women, who stand at the fulcrum of production and reproduction, have been disproportionately affected through:
> - Weakened participation in economic decision-making processes;
> - Diminished access to productive resources and services (e.g., land, credit and technology);
> - Erosion of food sovereignty;
> - Declining wages and destruction of livelihoods;
> - Violation of economic, social and cultural rights (e.g., the right to avail health and education services); and
> - Intensification of socially-ascribed reproductive or caring responsibilities especially in light of the HIV and AIDS pandemic (e.g., caring for the ill and fetching water and fuel).
>
> - Excerpted from the African Women's Statement on PWE[10] coming from the African consultation on PWE (2007)

Similarly, the youth statements from the African, Latin American and Caribbean, Asia-Pacific and European consultations on PWE point out

---

[9] The Guatemala Declaration may be accessed at
http://www.oikoumene.org/en/resources/documents/wcc-programmes/public-witness-addressing-power-affirming-peace/poverty-wealth-and-ecology/neoliberal-paradigm/agape-consultation-guatemala-declaration.html.
[10] The African Women's Statement on PWE is available at
http://www.oikoumene.org/en/resources/documents/wcc-programmes/public-witness-addressing-power-affirming-peace/poverty-wealth-and-ecology/neoliberal-paradigm/african-womens-statement-on-poverty-wealth-and-ecology.html.

that young people are among the victims of neoliberal policies especially in the areas of employment and education (see Box 2).

> 2. Youth and neoliberal policies
>
> There are few employment opportunities, even for those who get the privilege to obtain university degrees. When we do find work, wages are low, working conditions are inhuman, there is no job security and there are limited benefits...We move from irregular employment or short-term contract to another. We leave the countryside and go to the cities to look for work...We migrate overseas in the tens of thousands in search of jobs and a better life. We are pushed into prostitution, the drug trade, crime, and human trafficking...We want to go to school, but the majority cannot afford it. When we can go to school, schools are overcrowded and it is hard to learn. Courses are designed to fill the needs of the profit-driven, private corporations; therefore we study to get work, not to better ourselves or society.
>
> - Excerpted from the Youth Declaration[11] coming from the Asia-Pacific consultation on PWE (2009)

Financialization, Inequality and Crises

Another critical factor contributing to global and national inequality has been the rapid "financialization" of the global economy, with the turnover in financial markets expanding from 15 times world GDP in 1990 to almost 70 times world GDP in 2007 (Dillon in Kennedy, ed. 2012). As observed by Rosario Guzman (in Peralta, ed., 2010), "tremendous wealth is being created outside the production sphere with increasing financialization... [involving] increasing profitability of speculative and unproductive financial activities." Marcos Arruda (in Brubaker and Mshana, ed. 2010) describes this phenomenon as

---

[11] The Youth Declaration is available at
http://www.oikoumene.org/en/resources/documents/wcc-programmes/public-witness-addressing-power-affirming-peace/poverty-wealth-and-ecology/neoliberal-paradigm/asia-pacific-youth-hearing-on-poverty-wealth-and-ecology.html.

"profiting without producing": while a billion people suffer in poverty, "others make fortunes without producing, just by speculating."[12]

The last three decades saw the dramatic deregulation and globalization of the financial sector. At the same time, national and global imbalances widened significantly: in the United States, falling median wages contributed to rising household indebtedness (Dillon in Kennedy, ed. 2012); while in China, the low share of wages to national income helped to boost domestic savings and capital account surpluses. These various elements helped to usher in the sharpest and deepest financial and economic crisis since the Great Depression. Although originating from the United States, the global financial and economic crisis of 2008-2009 pushed tens of millions of people into unemployment, homelessness, indebtedness and poverty worldwide, eroding the attainment of the Millennium Development Goals (MDGs).

In response to this unprecedented financial and economic meltdown, the WCC prepared a statement for the United Nations (UN) Follow-up International Conference on Financing for Development in Doha in November 2008, which observed that "…the present international financial system is not merely inefficient. It is a system based on injustice, whereby the global poor are essentially subsidizing the rich."[13] Letters were likewise issued to the UN Assembly as well as to the Group of 20 in April 2009, calling on the international community to "go beyond short term financial bailout actions and to seek long term transformation based on sound ethical and moral principles which will govern a new financial architecture."[14]

In June 2009, the WCC re-convened the Advisory Group on Economic Matters (AGEM) to scrutinize the workings of the current financial

---

[12] See *Justice, Not Greed* edited by Pamela Brubaker and Rogate Mshana (2010).
[13] The statement may be downloaded at
http://www.oikoumene.org/en/resources/documents/wcc-programmes/public-witness-addressing-power-affirming-peace/poverty-wealth-and-ecology/finance-speculation-debt/statement-on-the-doha-outcome-document.html.
[14] The letter is available at
http://www.oikoumene.org/en/resources/documents/general-secretary/messages-and-letters/27-03-09-letter-to-g20.html.

system and to prepare a proposal leading to a new international financial architecture embracing ethics and especially justice as a core component of the Christian message. The discussions of the AGEM 09 are captured in the book *Justice, Not Greed* (Brubaker and Mshana, ed. 2010). The AGEM 09 also helped to shape the landmark WCC central committee statement on "Just Finance and an Economy of Life," [15] issued in September 2009.

The WCC central committee statement on "Just Finance and an Economy of Life" appeals for an ethical, just and democratic international financial regime "grounded on a framework of common values: honesty, social justice, human dignity, mutual accountability and ecological sustainability" and that "account[s] for social and ecological risks in financial and economic calculation; reconnect[s] finance to the real economy; and set[s] clear limits to, as well as penalize[s], excessive and irresponsible actions based on greed." Furthermore, the statement came up with specific proposals for critical reforms in the financial and monetary system including, among others, regulating the movement of volatile, speculative capital flows through the implementation of a financial transaction tax, exploring the establishment of a new global reserve system, and developing indicators that better capture socio-economic and ecological progress than the money-centric GDP.

Building on the statement on "Just Finance and an Economy of Life," the WCC central committee issued the "Statement on the Current Financial and Economic Crisis with a Focus on Greece"[16] in September 2012. It asserts that: "There is no justice when those who had little part in generating the crisis pay the highest price for it." Moreover, "[i]t is immoral to demand austerity and debt repayment at human and social cost which falls unfairly on the weaker members of society." The

---

[15] The statement may be accessed at
http://www.oikoumene.org/en/resources/documents/central-committee/geneva-2009/reports-and-documents/report-on-public-issues/statement-on-just-finance-and-the-economy-of-life.html.

[16] The complete text is available at
http://www.oikoumene.org/en/resources/documents/central-committee/kolympari-2012/report-on-public-issues/viii-statement-on-the-financial-crisis.html.

statement reiterates the call for the application of a financial transaction tax as a tool that will enable governments to meet peoples' economic, social and cultural rights.

Together with the World Communion of Reformed Churches (WCRC) and the Council for World Mission (CWM) and in partnership with civil society, the WCC through the PWE project continues to contribute to the discussions on building an alternative international financial order that ensures equity between and within nations in all aspects of economic life as well as the democratic functioning of global financial institutions. In October 2012, the WCC, WRCR and CWM jointly convened the Global Ecumenical Conference on a New Financial and Economic Architecture in Sao Paolo to identify criteria and prepare an ecumenical plan for transformation, resulting in the Sao Paolo Statement, "International Financial Transformation for an Economy of Life."[17]

It reads:
> We lament the manner in which economic and financial legislation and controls are biased in favour of the wealthy…[and] call for a system of just legislation and controls that facilitate the redistribution of wealth and power for all of God's creation…We reject the explosion of monetization and the commodification of all of life and affirm a theology of grace which resists the neoliberal urge to reduce all of life to an exchange value (Rom. 3:24)…We reject an economy that is driven by debt and financialization in favour of an economy of forgiveness, caring and justice and declare that debt and speculation have reached their limits. We affirm the words of the Lord's Prayer in which we pray to have our own debt forgiven in the same manner as we forgive the debts of others (Matt. 6:12)…We reject the ideology of consumerism and affirm an economy of Manna, which provides sufficiently for all and negates the idea of greed

---

[17] The full text of the Sao Paolo Statement may be accessed at http://www.oikoumene.org/en/resources/documents/wcc-programmes/public-witness-addressing-power-affirming-peace/poverty-wealth-and-ecology/finance-speculation-debt/sao-paulo-statement-international-financial-transformation-for-the-economy-of-life.html.

(Ex. 16) ... We reject increasing individualistic consumerism by affirming and celebrating the diversity and interconnectedness of life. We further affirm that wholeness of life can be achieved only through the interdependent relationships with the whole of the created order...

The Sao Paolo Statement calls for the creation of a UN Economic, Social and Ecological Security Council, where pressing economic, social and ecological issues would be brought together to be discussed and acted upon in a coherent way, as well as an International Monetary Organization (IMO) to replace the International Monetary Fund, which would have oversight over monetary policies and would deploy funds without structural adjustment conditions to establish an effective, stable, fair and socially responsible global financial and economic architecture.

Ecological Debt

Though today's world is a more prosperous place (in monetary terms), such prosperity has come at a cost to ecological health as evidenced by the confluence of environmental problems confronting our world today: deforestation, climate change, ocean acidification, biodiversity loss, and so on. Wealth creation generates tremendous ecological debts.[18] Ecological debt lenses expose the fact that it is the global South who is the principal ecological creditor while the global North is the principal ecological debtor.

The ecological debt of the latter arises from several causal mechanisms: through loan conditionalities as well as multilateral and bilateral trade and investment agreements that pressure poor countries to pursue export-oriented and resource-intensive growth strategies that ignore the costs of erosion of ecosystems and increasing pollution; and through development projects in the South financed by international financial institutions in collusion with undemocratic governments, without the informed consent of local communities and with minimal thought for the projects' ecological and social consequences.

---

[18] For an introduction to ecological debt, see *Ecological Debt: The Peoples of the South are the Creditors,* edited by Athena Peralta (2006), which can be downloaded at http://www.deudaecologica.org/publicaciones/CoverTablePreface.pdf.

As both the Dar es Salaam Statement and the Guatemala Declaration emanating from the African and Latin American and Caribbean consultations on PWE emphasize, ecological debt also arises from unsustainable consumption patterns (see Box 3) as well as from ecologically-damaging production methods. In particular, the Dar es Salaam Statement on PWE [19] challenges churches in the North to "acknowledge the privileges derived from complicity – through their production and consumption patterns – in systems of domination and exploitation that dehumanize and destroy life in Africa."

---

3. Consumption patterns in the United States

Between 1975 and 2000, total material consumption in the U.S. grew by 57 percent. U.S. consumers are not only buying more goods but also eating greater amounts and larger portions of food. U.S. food consumption increased 16 percent since 1970. According to the U.S. Environmental Protection Agency, 42 percent of U.S. greenhouse emissions result from the production, consumption, and disposal of food. People in the U.S. use an average of 134 more gallons of water per day (about 159 gallons) than the half of the world's population that lives on 25 gallons of water per day.

-Excerpted from "As Any Might Have Need"
(Hinson-Hasty in Kennedy, ed. 2012)

In "Wealth Creation, Poverty and Ecology in Africa," Clement Kwayu (in Mshana, ed. 2012) discusses the ecological debt accrued by oil production in Nigeria, which has attracted billion-dollar investments by multinational corporations such as Exxon Mobile, Chevron Texaco and Conoco Phillips, but has done little to alleviate poverty in the country. Between 1986 and 2000 alone, the Niger Delta was affected by 3,854 oil spills that released 437,810 barrels of oil into the Nigerian environment.

---

[19] The Dar es Salaam Statement may be accessed at
http://www.oikoumene.org/en/resources/documents/wcc-programmes/public-witness-addressing-power-affirming-peace/poverty-wealth-and-ecology/neoliberal-paradigm/dar-es-salaam-statement-on-linking-poverty-wealth-and-ecology.html.

The oil spills caused massive destruction of farmlands, polluted sources of drinking water, wiped out mangrove forests and fishing grounds, and decimated colonies of fishes, crustaceans and birds. People living in the area have reported health problems stemming from the lack of safe drinking water.

Similarly, John Dillon (in Kennedy, ed. 2012) draws attention to Canada's petroleum industry, particularly tar sands development in the province of Alberta, which has played a substantial role in wealth generation, accounting for 11.7 percent of all industry profits from 2000-2009, but at a high price to ecological health (see Box 4).

---

4. Ecological debt and the tar sands

The ecological destruction wrought by the extraction of bitumen from the tar sands begins with the removal of boreal forest. While 686 square kilometres of boreal forest have been removed so far to enable mining of the tar sands, the total could rise to 4,800 square kilometres if mining operations are allowed to expand.

Where bitumen is extracted through the injection of steam into deeper sand formations, the construction of roads and clearing of well sites also disrupts the habitat for caribou, moose and other animals on which Indigenous Peoples depend for sustenance. Scientists predict that the woodland caribou herd could disappear from the north east of Alberta. Chief Janvier of the Chipewyan Prairie Dene First Nation laments: "The extinction of caribou would mean the extinction of our people. The caribou is our sacred animal; it is a measure of our way of life. When the caribou are dying, the land is dying."

Removal of boreal forest also impacts climate change by eliminating the forest's capacity to act as a carbon sink. The extraction of bitumen and its upgrading into synthetic fuel ready for refining emit 3.2 to 4.5 times more greenhouse gas (GHG) than conventional oil extraction.

-Excerpted from "Poverty, Wealth and Ecology in Canada" (Dillon in Kennedy, ed. 2012)

Climate change is perhaps one of the most striking examples of ecological debt, highlighting economic, social and ecological inequities between the global North and global South. While rich, industrialized countries are mainly responsible for GHG emissions causing climate change (though large, emerging economies are becoming major contributors to global GHG emissions in absolute terms), research indicates that poor countries will assume a bigger burden of the adverse effects of a warming climate including: the displacement of people living in low-lying coastal areas and small island states; the loss of sources of livelihood, food insecurity, reduced access to water, and forced migration. At the same time, poor countries have the least resources, such as capital and technology, to adapt to climate change.

Ultimately, however, climate change could put all life at stake. Thus the Budapest Call issued by the European consultation on PWE[20] considers climate change "the greatest threat to the future of our planet" and demands that "climate justice...be realized between people, countries and generations, humans and non-humans and with the Earth itself" (see Box 5).

---

5. Climate justice

Climate justice requires social justice. Climate justice includes the right to development, particularly in weaker economies. Climate justice requires the development of renewable energy and economies of sufficiency inspired by an ethic of self-limitation. Climate justice is a condition for the eradication of poverty and the eradication of poverty is a condition for climate justice. Climate justice demands the primacy of democratic politics over economics and the embedding of market economies in social and cultural contexts.

- Excerpted from the Budapest Call for Climate Justice
coming from the European consultation on PWE (2007)

---

[20] The Budapest Call may be accessed at
http://www.oikoumene.org/en/resources/documents/wcc-programmes/public-witness-addressing-power-affirming-peace/poverty-wealth-and-ecology/neoliberal-paradigm/agape-consultation-budapest-call-for-climate-justice.html.

Against this background, and following a consultative process among member churches, the WCC central committee issued a statement on "Eco-justice and Ecological Debt"[21] in September 2009. Among others, the statement calls on Northern governments, institutions and corporations to take initiatives to drastically reduce their GHG emissions within and beyond the UN Framework Convention on Climate Change; urges the international community to ensure the transfer of financial resources to countries of the South to keep petroleum in the ground in fragile environments, to preserve other natural resources as well as to pay for the costs of climate change mitigation and adaptation; and encourages churches to develop campaigns on ecological debt and climate change as well as build awareness and deepen theological reflection on a new cosmological vision of life, eco-justice and ecological debt through study and action.

To follow-up the WCC Statement on "Eco-justice and Ecological Debt," the WCC has collaborated with the Economic Justice Network of the Fellowship of Christian Churches in Southern Africa (FOCCISA), United Church of Canada, Kairos Canada, Church of Sweden, Jubilee South, Accion Ecologica in Ecuador, Integrated Rural Development of the Weaker Sections in India and Panalipdan Philippines in organizing a wide range of activities to expand awareness of ecological debt, climate change and resource-extractive industries, to establish support for a people's tribunal on ecological debt, as well as to begin to collect case studies toward demanding ecological debt reparations.

Greed in the Economy

At the height of the global financial crisis, the WCC in a letter to the G20 noted that "greed has become the basis for growth." Today, there is wide acceptance, even endorsement of greed, reflecting a change in moral and cultural thinking that is linked to the neoliberal revolution that has transpired over the last 30 years. Greed has become officially sanctioned in our economic systems which have as inherent goals:

---

[21] The statement is available at
http://www.oikoumene.org/en/resources/documents/central-committee/geneva-2009/reports-and-documents/report-on-public-issues/statement-on-eco-justice-and-ecological-debt.html.

limitless growth, the generation of wealth and the highest possible returns in the shortest time frame, and the maximization of utility or pleasure from the consumption of material goods.

The WCC-commissioned studies on the links between poverty, wealth and ecology in Africa (Mshana, ed. 2012), Asia (Peralta, ed. 2010), Latin America and the Caribbean (Mshana, ed. 2009), Europe (Pavlovic, ed. 2011) and North America (Kennedy, ed. 2012) provide substantiation that wealth creation has often taken place at the expense of people's well-being and the health of ecosystems. These findings support the contention that: "Only by defining the upper limits of consumption and thus of income for the rich can the real needs of the poor be satisfied and the impact of the economy on the environment be brought under control" (De Lange and Goudzwaard 1995).

It was, however, the APRODEV study project on *Christianity, Poverty and Wealth* (Taylor, ed. 2003) that first proposed the development of wealth or greed lines to stand in counterpart to poverty lines:

> Can excessive wealth be defined as concretely as we sometimes define poverty? Is there a wealth line above which no one should rise just as there is a poverty line below which no one should be allowed to fall? Can we speak of "relative wealth" in the way we speak of "relative poverty" so focusing once again on the unacceptable disparities within communities and countries, rich or poor, as well as between them? What might be the indicators of excessive wealth to stand alongside poverty indicators…?

Reacting to these challenges, the WCC brought together the Greed Line Study Group, which met thrice from 2009 to 2011, to establish how greed could be measured and monitored; to work out ethical, theological and moral guidelines for just and sustainable production and consumption; and to come up with proposals to avert greed in our economic systems. Some of the main findings of the group are discussed in the succeeding paragraphs.

As a first step, greed could be defined as the desire to have more than one's legitimate share of wealth and power (Raiser 2011). In a world dominated by modern capitalist thinking, there is a need to demonstrate through objective analyses that greed is a highly damaging form of desire with negative consequences not only for vulnerable peoples and communities, but also for our increasingly fragile ecosystems. The impact of greed on ecology has gained greater urgency in the last decade as the effects of climate change are already being felt.

Greed is driven by complex dynamics; it has to be interrogated individually, institutionally, structurally and culturally. More attention, however, ought to be directed at the kind of greed that is implanted and often concealed in socio-economic policies, institutions, structures as well as in the prevailing ideologies of our times which legitimate and propagate a culture of greed. Today, there exist structural arrangements which facilitate, foster, demand and even presuppose the development and expression of greedy desires on the part of individuals or social groups, so that one could speak of "structural greed" or "institutionalized greed" with structural consequences. This demands counter-measures at the structural level. However, in illuminating the workings of greed, it is not sufficient to analyze "structural greed" or "institutionalized greed." It is increasingly apparent that the former works in tandem with a "culture of greed" or "habitual greed" that is shaping collective thinking and behaviour.

The key functions of a greed line and greed indicators are to expose the collective and structural manifestations of greed and their economic, social and ecological consequences; and to serve as "red lights" that signal to both citizens and policymakers that critical limits are being overstepped. They are tools for awareness-building, advocacy and critical discernment.

In this connection, Carlos Larrea (2011) develops an ethical social consumption function that factors in inequitable socio-economic conditions and ecological limits, and derives the greed line from the said function. The greed line is defined as "the maximum ethically acceptable individual consumption in the current global economy, simultaneously characterized by the social exclusion of a large proportion of the world population affected by poverty, a high level of national and international inequality, luxurious consumption accounting for a large share of global output, and the lack of environmental sustainability, affecting the right of future generations to achieve adequate living conditions."

Apart from developing a greed line, greed could be measured and monitored by identifying a strategic set of multidimensional greed indicators which have the objective of alerting policymakers and the general public to critical manifestations of greed at collective and structural levels and which could inform the development and implementation of policies and measures to avert greed and plan equitable and sustainable production, consumption and distribution.

The provisional report of the Greed Line Study Group entitled "Churches Addressing Greed in the Economy" calls on churches to continue to improve and monitor indicators and indices of greed together with civil society; to engage in critical self-reflection and transformation of churches' own production, consumption and investment behaviour; to advocate for anti-greed measures at the national and global level such as those related to progressive taxation, interest rate ceilings, and fair trade; to promote cultural values in support of the "economy of enough" including through campaigns for simple living; and to advance eco-justice.

Social Justice and Common Goods

In the current era of globalization, profit orientation and market rivalry have increasing dominion over the equitable sharing of goods and resources, depriving many people of the same. In response to this concern, the WCC's Commission of Churches in International Affairs which comprises PWE issues released a policy paper on "Social Justice and Common Goods" [22] in 2011.

---

[22] The policy paper may be accessed at
http://www.oikoumene.org/en/resources/documents/wcc-commissions/international-affairs/economic-justice/social-justice-and-common-goods-policy-paper.html.

Common goods could be defined as all that is essential for a life in fullness and dignity for all God's creation. These include: land, water, air, forests, health, education, shelter, energy, transport, peace, human security, information, knowledge, solidarity and freedom. When people enjoy universal access to common goods there is social justice.

The policy paper notes that:

> Common goods are under a growing threat of commodification resulting in climate change, poverty and inequality. The power of transnational corporations is setting the geopolitical and geo-economic scene. Their power gives them a stronghold on the rules of trade and finance and exacerbates existing inequalities between rich and poor. States are losing their capacity to fulfill their main functions of regulating economy, protecting the environment, defending social cohesion and values and guaranteeing their peoples' security.

Further, the policy paper calls on churches to raise the concern over the tendency of reduction of state commitment and involvement in managing of public utilities, and the trend toward decreasing social protection; to collect and share stories from people experiencing the adverse effects of the commodification of common goods; to organize debates and conversations with at the grassroots level as well as will economic and political leaders; to carry out awareness raising and advocacy campaigns; and to help build a global movement for justice and human rights.

Economic and Ecological Injustice and Violence

Massive socio-economic disparities lie at the centre of expanding ripples of social and political upheaval including democratic uprisings in the Middle East and the "Occupy" protests flaring up across many cities in Europe, United States and other parts of the world since 2011. The broad experience of material deprivation and insecurity amid the unimaginable prosperity enjoyed by a handful erodes social cohesion and generates social and political unrest – which has often been met by brute police and military force to secure the political and economic interests of a powerful elite.

The regional consultations on PWE carefully considered the connections between economic and ecological injustice and violence. In particular, the "Chiang Mai Declaration on PWE"[23] points out:

> In Asia and Oceania as in elsewhere, violence has often been used by the economically and politically powerful in securing the planet's "natural resources." Imperialist terror and greed desecrate both Mother Earth and women's bodies. We listened with heavy hearts to stories of church people gunned down in the Philippines for defending ecology and farmers' and workers' rights; communities dying from toxic pollutants in military bases; intensified violence against women in militarized zones and in their own homes.

Similarly, the African Women's Statement on PWE observes:

> The intensifying competition among powerful nations and multinational corporations for Africa's oil, minerals and lumber has not only deprived African people of the use of these resources for their well-being. In the scramble to secure these resources through political pressure, military and paramilitary force, wars and conflicts have erupted in the continent, accompanied by massive human rights violations. The growing nexus between neo-liberal economic globalization and militarization is a reflection of the face of empire in Africa.

Thus at the International Ecumenical Peace Convocation (IEPC) held in Kingston in May 2011, churches affirmed that measures to close socio-economic gaps and to address ecological debt are important pathways to strengthening social cohesion and building lasting peace at local, national and global levels. In particular the message coming from the IEPC urges governments "to take immediate action to redirect their financial resources to programmes that foster life rather than death" and "to

---

[23] The Chiang Mai Declaration may be accessed at
http://www.oikoumene.org/en/resources/documents/wcc-programmes/public-witness-addressing-power-affirming-peace/poverty-wealth-and-ecology/neoliberal-paradigm/agape-consultation-chiang-mai-declaration.html.

reconstruct radically all our economic activities toward the goal of an ecologically sustainable economy."[24] It also encourages churches to adopt common strategies that "address more effectively irresponsible concentration of power and wealth as well as the disease of corruption... includ[ing] more effective rules for the financial market, the introduction of taxes on financial transactions and just trade relationships."

---

[24] The IEPC message is available at http://www.overcomingviolence.org/en/resources-dov/wcc-resources/documents/presentations-speeches-messages/iepc-message.html.

3.

Issues of Faith: Theological Reflection

Theological Responses to Issues of Poverty, Wealth and Ecology

Recognizing that the interwoven crises confronting our world today are, at core, ethical, moral and spiritual, the PWE project offered spaces for churches to reflect jointly and critically on the theological implications of and responses to rampant poverty, widening wealth disparities and ecological degradation through regional and global church consultations as well as various seminars and study processes. Some of the questions that were tackled include:

- What are the ethical, moral and spiritual implications of the current intertwined financial, economic and ecological crises?
- What does faith have to say about greed?
- In thinking about an "economy of enough" that is respectful of ecology what lessons can be drawn from other faiths and the radical spirituality and contentment of many indigenous and rural communities?

Reflecting on Poverty:
God's Preferential Option for the Poor

Highlighted in almost all of the messages arising from the regional consultations on PWE, God's preferential option for the poor serves as both primary motivating force as well as a crucial touchstone for churches' response to economic matters, from issues of domestic taxation to issues of international trade. Likewise, the deep concern expressed by churches over the recent financial and economic turmoil stems largely from the incredibly heavy impacts the crisis has had and continues to have on the most vulnerable, the excluded and the poor – the very people that God calls us to especially defend and care for.

Numerous biblical texts in both Old and New Testaments – from the book of Isaiah ("Learn to do good; seek justice, rescue the oppressed, defend the orphan, plead for the widow," Isaiah 1:17) to the Gospel of Matthew ("For I was hungry and you gave me something to eat, I was thirsty and you gave me something to drink, I was a stranger and you invited me in, I needed clothes and you clothed me, I was sick and you looked after me, I was in prison and you came to visit me… What you did to one of the least of these, who belong to me, you did for me" Matthew 25:34) – demonstrate over and over again that care for the situation of the poor and all efforts to improve their condition and address the systemic roots of poverty form a central and indispensable pillar of the Christian faith.

Thus the 2012 AGAPE Call to Action issued from the Global Forum on PWE entitled "Economy of Life, Justice and Peace for All"[1] (see Economy of Life, 41ff) strongly proclaims:

> Our vision of justice is rooted in God's self-revelation in Jesus Christ who drove money changers from the temple (Matthew 21:12), made the weak strong and strong weak (1 Corinthians 1:25-28), and redefined views of poverty and wealth (2 Corinthians 8:9). Jesus identified himself with the marginalized and excluded people not only out of compassion, but because their lives testified to the sinfulness of the systems and structures. Our faith compels us to seek justice, to witness to the presence of God and to be part of the lives and struggles of people made weak and vulnerable by structures and cultures – women, children, people living in poverty in both urban and rural areas, Indigenous Peoples, racially oppressed communities, people with disabilities, Dalits, forced migrant workers, refugees and religious ethnic minorities.

---

[1] The 2012 AGAPE Call to Action is available at
http://www.oikoumene.org/en/resources/documents/wcc-programmes/public-witness-addressing-power-affirming-peace/poverty-wealth-and-ecology/neoliberal-paradigm/agape-call-for-action-2012.html.

Reflecting on Wealth:
Social Responsibility and Overcoming Greed

While churches have for a long time ruminated about poverty, the issue of wealth – its origins and uses – has, at least until recently, received scant theological consideration. However, the 2008 financial implosion drew public as well as churches' attention to the role played by the desire for unlimited wealth accumulation – in a word: greed – in the economy.

The biblical tradition, in fact, provides a consistent normative framework for the social responsibility of wealth grounded on the conviction that God is the creator who lovingly provides for all living beings what they need in order to live and live fully. As wealth is a blessing from God (rather than the product of human effort), "it is to be used for the benefit of the whole community and especially for those who are unable to provide for their own needs" (Raiser 2011). Moreover, the Decalogue clearly rebukes the desire to enlarge one's wealth at the expense of other people.

Perhaps one of the best known biblical passages on greed is the one in the book of Luke (12, 15) which cautions: "Take care! Be on your guard against all kinds of greed; for one's life does not consist in the abundance of possessions." However,
greed is not only about the desire for unlimited accumulation of money and resources. Konrad Raiser (2011) sums up the biblical injunction against greed as follows:

> The dynamic of greed is not limited to material possessions; it is an expression of the thirst for power, the temptation to outdo or take advantage of others (1 Thessalonians 4:4ff) and can thus become a prime example of vice and lawlessness (Romans 1:29). The condemnation of greed is not only a moral judgment; rather, it constitutes an act of idolatry by focusing trust on material possessions or on one's power rather than on God (Ephesians 4:19; 5:5). Greed is considered as a sign of loss of faith (1 Timothy 6:10) and a dangerous temptation for those in positions of authority (1 Thessalonians 2:5; 2 Corinthians 9:5). Greed is not only idolatry but essentially a denial of Christ who "did not regard

equality with God as something to be exploited, but emptied himself…" (Philippians 2:6f; also Matthew 25:31ff). Therefore, the Christian community is being admonished to "do nothing from selfish ambition or conceit" and to look "not to your own interests, but to the interests of others" (Philippians 2:3f).

In overcoming greed, theology and churches have an essential role to play in exposing the falsehoods disseminated by the dominant economic order and worldview: that human action can achieve perfection and the infinite and that greater consumption and economic success lead to complete self-fulfillment. Jung Mo Sung (2011) argues that: "Christian churches should witness to a spirituality that accepts the human condition and knows that our "being" does not consist in possessions, in products, in grand mansions or imposing churches, but rather in loving relationships between people."

Moreover, where people fail to grasp the structural links between their desire for a higher standard of living and the poverty endured by their neighbours, churches have the prophetic task of making visible and lifting up the voices of the victims of the economy of greed. In addition, they have the pastoral responsibility to enable people who are captive to a spirituality of consumerism to recognize the painful consequences of their way of living on other people and on the environment.

In this connection, the identification of a greed line or a set of indicators of greed could serve as guidance to churches and Christians. According to Konrad Raiser (2011):

> Acknowledgment of a "greed line"…requires a fundamental metanoia, i.e., the spiritual recognition that the fullness of life can only be experienced as a gift that is shared in community. Indeed, life, freedom, power, and love lose their value once they are claimed and defended as individual property or right; they increase and gain in strength as they are being shared. The acknowledgment of a "greed line" is therefore an act of "spiritual discernment," i.e., of unmasking the temptations of the false spiritualities of unlimited accumulation and consumerism. The decisive

criteria should be derived from an assessment of what sustains or undermines and destroys life in just and sustainable relationships in human community and with the natural world.

Reflecting on Ecology: Care for Creation

The Earth and all who live in it are part of God's beautiful creation (Psalm 24). In their environmental advocacy, churches have often used the concept of stewardship to emphasize humanity's responsibility for protecting and preserving ecosystems from which people, especially the poorest, draw their daily sustenance. In Psalm 148, God invites us to care for creation, a sacred gift from God, so that the whole creation rejoices in harmony and praise.

However, indigenous contributions to the regional consultations on PWE point out that the notion of stewardship may still be one that is anthropocentric. According to the Indigenous Peoples' Statement from the Latin American and Caribbean consultation on PWE, [2] the fundamental centre in the indigenous Cosmo vision "is not the human being but creation and […] human beings form part of creation and […] everything that forms part of creation is important and is in a permanent search for harmony and balance." On the other hand, the Western interpretation of Christianity tends to separate humanity from the rest of creation. It also "separates the religious and the spiritual from the social, economic and political" and "promot[es] an individualistic vision that encourages egoism…and does not take into account the community and the collective vision of our Indigenous Peoples."

Toward an Economy of Life

Churches' witnessing for justice in the economy and the Earth is founded on the biblical vision of fullness of life for all: God created the household of life (*oikos*), humans and non-human beings to live in community with one another (Psalm 115:16 and Genesis 1-2).

---

[2] The Latin America and Caribbean Indigenous People's Statement is available at http://www.oikoumene.org/en/resources/documents/wcc-programmes/public-witness-addressing-power-affirming-peace/poverty-wealth-and-ecology/neoliberal-paradigm/agape-consultation-guatemala-indigenous-peoples-statement.html.

According to the Dar es Salaam Statement:

> The Christian notion of oikos resonates with the African understanding of ubuntu/botho/uzima (life in wholeness) and ujamaa (life in community). They embrace among others, the values of fullness of life, full participation in all life processes including in the economy and ecology. It further entails the just care, use, sharing and distribution of resources and elements of life. Where the above and life-affirming relationships have been violated, the institution of restorative, redistributive and rectificatory (wisdom) justice are necessary. These principles of justice, reparation, restoration and reconciliation, forgiveness, mutual love and dignity for all of God's Creation ought to be promoted ecumenically as bases for constructive critique of global capitalism, which increasingly violates life-in-abundance (John 10:10).

As noted by the Latin American and Caribbean Indigenous Statement, the biblical concept of life-in-abundance finds its parallel in the Cosmo vision of Indigenous Peoples that celebrates the "good life" (*Suma Jakana* for the Aimara of Bolivia, *Sumak Kawsay* for the Kichua of Ecuador, *Ut a Wach* for the Quiche of Guatemala, *lekil K'uxlejal* for the Tseltal of Mexico, *aoya poh laka* for the Miskitu of Nicaragua): "It is good to live here and now…Our Indigenous Peoples want not only spiritual salvation and political, economic and social development, but also a good life, life in abundance, in an all-embracing way."

Action is an extension of faith (James 2:4). In this light, the Chiang Mai Declaration issued by the Asia-Pacific consultation on PWE celebrates "the spirituality that is found in resistance and political engagement" and calls for the building of alternative economies as part of God's work:

> Genuine faith and spirituality entail action. We assert that the multiple crises we confront today urgently demand radical and collective responses, not only from Asia and Oceania, but also from the worldwide faith community. United in God's love, we can and must begin to construct flourishing and harmonious economies where:

- all participate and have a voice in the decisions that impact on their lives;
- people's basic needs are provided for through just livelihoods;
- social reproduction and the care work done predominantly by women are supported and valued; and
- air, water, land and energy sources that are necessary to sustain life are protected and preserved.

In short, we can and must shape Economies of Life and Economies for Life.

Against this background, the 2012 AGAPE Call to Action on "Economy of Life, Justice and Peace for All" asserts:

God…affirmed the goodness of the whole creation (Genesis 1)…The whole community of living organisms that grows and flourishes is an expression of God's will and works together to bring life from and give life to the land, to connect one generation to the next, and to sustain the abundance and diversity of God's household (oikos). Economy in God's household emerges from God's gracious offering of abundant life for all (John 10:10)."

Toward Eco-Justice

The intertwined economic and ecological emergency that we face today demands that churches take a radical perspective that advances eco-justice. The WCC Statement on "Eco-justice and Ecological Debt" observes:

Churches have been complicit in this history [of ecological devastation] through their own consumption patterns and through perpetuating a theology of human rule over the Earth. The Christian perspective that has valued humanity over the rest of creation has served to justify the exploitation of parts of the Earth community. Yet, human existence is utterly dependent on a healthy functioning Earth system.

Humanity cannot manage creation. Humanity can only manage their own behaviour to keep it within the bounds of Earth's sustenance. Both the human population and the human economy cannot grow much more without irreversibly endangering the survival of other life forms. Such a radical view calls for a theology of humility and a commitment on the part of the churches to learn from environmental ethics and faith traditions that have a deeper sense of an inclusive community.

The Bible – particularly in the books of Jeremiah (14:2-7), Isaiah (23:2-7) and Revelation (22) – illuminates the intrinsic connection between socio-economic injustice and ecological crises. As the Calgary Call issued by the North American consultation on PWE[3] asserts: "The cry of the Earth and the cry of the poor are one." Furthermore, the Budapest Call from the European consultation on PWE comprehends that humanity's salvation lies also in "the restoration of a broken relationship with the whole created order."

Eco-justice brings together economic, social and ecological justice, recognizing that economy and ecology are inherently interconnected and that the objective of poverty eradication cannot be attained independently of ecological sustainability. In this vein, the WCC Statement on "Eco-justice and Ecological Debt" calls on churches 'to proclaim God's love for the whole world and to denounce the philosophy of domination that threatens the manifestation of God's love." Based on Jesus' greatest commandments "to love God with all of your heart" and "to love your neighbour as you love yourself" (Luke 10:27), churches "must broaden its understanding of justice and the boundaries of who our neighbours are."

---

[3] The Calgary Call may be accessed at www.oikoumene.org.

4.

Dialogue on Economic Globalization: Religious, Economic and Political Actors

The Importance of Dialogue in an Era of Globalization

As pointed out in part 2 of this report, the impacts of economic globalization have been uneven, adversely affecting some regions, communities and social groups – often those who are socio-economically weak to begin with – while benefiting others that have the know-how, technology and capital to take advantage of the opportunities generated by liberalized markets. Moreover, churches and faith-based organizations do not necessarily have the same interests as businesses and governments. Consequently, there are differences in analyses and responses among churches, and between religions and other economic and political actors. At the same time, transforming the economic system toward eradicating poverty and ensuring ecological sustainability entails speaking truth to power as well as requiring coordinated responses in all levels from a broad spectrum of stakeholders – which is only possible where there is some common understanding or vision. From this perspective, then, the importance of dialogue in order to build mutual understanding and stimulate collaboration with like-minded stakeholders cannot be overstressed.

The PWE project has worked according to the insight that churches have a special role to play in creating deliberate openings for strengthening dialogue and building synergies arising from contrasting positions. Moreover, according to Bishop Julio Murray (in Pavlovic, ed. 2011): "Sometimes the Church does not need to be the voice of those who do not have a voice. Sometimes the Church needs to be the body that articulates the idea and makes sure that the voice of the most vulnerable can be heard."

Thus the regional consultations on PWE have usually featured: (1) an immersion programme to enable dialogue and engagement with local communities and their realities; (2) a panel of representatives of churches from other regions in order to facilitate North-South exchanges and discussion; (3) hearings of women, youth, Indigenous Peoples and migrants, which aim to lift up the voices of sectors in society that are usually marginalized; as well as (4) spaces where civil society, business, government and church perspectives could be brought to bear as well as be in discourse with each other on issues of poverty eradication, wealth redistribution and ecological protection (see Box 6).

---

6. Dialogue between state, business and grassroots sectors during the consultation on PWE in Africa

The African consultation on PWE held in Dar es Salaam in November 2007 was set off by keynote presentations from the Office of the Prime Minister of Tanzania, a wealthy Tanzanian business man and a Tanzanian woman farmer. The presentations elaborated on the linkages between poverty, wealth and ecology from contrasting social locations. The state perspective highlighted the role of global trade and financial systems, particularly agricultural trade policies and external debt instruments, in impoverishing Africa. The corporate perspective conceded that profit-oriented wealth creation often results in exploitation of workers and environmental damages, but offered examples of how these problems could be redressed in small ways. Hearing the side of business was particularly illuminating in terms of understanding the motives behind wealth creation and accumulation. From the perspective of a poor woman farmer, states and corporations were challenged to act decisively on corruption and on growing inequalities between classes and genders, including through education programmes.

The succeeding paragraphs provide concrete and inspiring examples of dialogue and engagement among churches and between churches and other agents within society with a focus on highlighting key outcomes and lessons learned.

Dialogue between European and Latin American Churches on Globalization

As a joint contribution to the PWE process, the Conference of European Churches (CEC) and the Latin American Council of Churches (CLAI) embarked on a dialogue on "The Threats and Challenges of Globalization" from 2009 to 2010 with the objective of overcoming distrust, prejudices and lack of information on both sides.[1]

The first part of the dialogue covered areas where the churches of CEC and CLAI expressed differences in opinion. These include the role of the state, the role of the market and the function of empire. It is observed that: "Dialogue, although intensive and sincere, does not always and necessarily eliminate differences. The different histories of Europe and Latin America, different socio-political set-ups and different impacts, in form and intensity, of current globalization processes are the causes of different experiences and standpoints, sometimes using the same language in different ways" (Pavlovic, ed. 2011).

The second part of the dialogue addressed "signals of change" or the grounds where both European and Latin American churches have discovered strong commonalities which could form the base for common action. Both CEC and CLAI "plea for a stronger interdependence between politics, economics and civil society and for a strengthening of the effort toward a sufficiency economy, which is set as a counter-image to an economy based on greed and financed through extensive and ever-increasing debts" (Pavlovic, ed. 2011).

The last part of the dialogue identified areas for possible joint action between churches of both continents, namely: climate justice, ecological debt, illegitimate debts, hunger and food crises, and water as a global challenge and a human right.

---

[1] The outcomes of the CEC-CLAI dialogue on globalization may be accessed in both English and Spanish at
http://ceceurope.org/fileadmin/filer/csc/Economic_Globalisation/Amenazas_Ingles.pdf

In outlining the way forward, CEC and CLAI raised the following questions: "Can we define together joint programmes empowering civil society, to think anew the role of the Church in a prophetic and contextual way? Can we create a strategic alliance between CEC and CLAI?"

The dialogue concluded with a commitment from both CEC and CLAI to "resolutely work toward a just and fair international economic order in which no one has to starve and all can live a life in dignity and fullness."

Interfaith Dialogue on Poverty, Wealth and Ecology as well as on Structural Greed

The Asia-Pacific consultation on PWE convened an interfaith panel on poverty, wealth and ecology, recognizing that, especially in the context of Asia, Christians are a religious minority. The discussion helped to shape the Chiang Mai Declaration emanating from the aforementioned conference, which states:

> From other ancient faiths and religions birthed in Asia, we learned about Buddhism's "middle way;" Hinduism's ahimsa (nonviolence) toward ecology and all human beings; and Islam's injunction to fight oppression in all its forms.

The consultation recommended the conduct of deeper dialogue on poverty, wealth and ecology with multi-faith communities to bring about more meaningful solidarity.

The WCC's PWE project also contributed to a series of interfaith discussions from 2010 to 2011 on the issue of structural greed initiated by the Lutheran World Federation's department on theology and public witness as well as the WCC's programme on interreligious dialogue and cooperation.

The Buddhist-Christian dialogue revealed that both traditions denounce greed as a vice that is manifested in individual, social and structural levels. According to the "Buddhist-Christian Common Word on

Structural Greed,"[2] "to avoid addressing structural greed and to focus on individual greed is to maintain the status quo." It also points out that:

> Self-interest, necessary for human well-being, does not necessarily constitute greed. Insofar as humans can survive and flourish only together with one another, self-interest naturally includes the interests of others. Therefore, when self-interest is pursued without compassion for others, when interconnectedness is disregarded or when the mutuality of all humanity is forgotten, greed results.

During the dialogue, Buddhists and Christians agreed that strategies for addressing greed at the personal and social levels include promoting generosity and cultivating compassion for others through effective preaching and teaching as well as spiritual practices such as meditation and prayer; while thwarting greed embedded in political and economic structures requires the institution of anti-greed measures that regulate financial transactions and promote the equitable distribution of wealth.

On the other hand, the conference on "Muslims and Christians Engaging Structural Greed Today" helped to expose areas of common understanding around the following concepts: economy; scarcity, abundance, and stewardship; social, cultural, and ecological commons; and trust.[3] In particular, both Muslims and Christians concur that:

> Greed as a form of structural impoverishment and social depravity is an impediment to the generous giving that should

---

[2] The "Buddhist Christian Common Word on Structural Greed" is available at www.oikoumene.org.
[3] The conference findings on "Muslims and Christians Engaging Structural Greed Today" may be downloaded at
 http://www.globethics.net/web/the-lutheran-world-federation-dn/collection-articles?p_auth=eF856eFv&p_p_id=generic_search_list_portlet_WAR_digitallibraryspring25portlet&p_p_lifecycle=1&p_p_state=normal&p_p_mode=view&p_p_col_id=column-2&p_p_col_count=1&_generic_search_list_portlet_WAR_digitallibraryspring25portlet_action=select&function=showMetadata&tableName=lucene&docId=5150025

define human economic activity. Greed is a form of debilitation whenever it ruptures the common good in favour of personal interest. Systemic structures of greed are grounded in this rupture, so that greed is understood to be a virtue, and generosity a naïve value of the lesser equipped. But this rupture and reversion is contrary to the shared core of the Muslim and Christian value of the human being in relation to God and society.

The conference findings concluded with a recognition of the challenges posed by neoliberal economic globalization to people as well as communities of faith and a call to joint action:

> Muslims and Christians must resist and reverse the privatization of the commons. It follows that our collective response to these challenges must be interreligious in nature, drawing strength from the rich heritage of each religion as much as from the shared values of both religions. In other words, since the problem of structural greed does not distinguish between different religions, our resistance must transcend and rise above religious differences.

As part of its recommendations to churches, the Greed Line Study Group considers a deeper Muslim-Christian discussion on usury as particularly important inasmuch as the prohibition on charging interest on loans is common to both Qur'anic and biblical teachings. The Christian interpretation of biblical injunctions against usury has adapted and continues to adapt to a modern, capital-based and increasingly globalized economy.

The European Union, China and Africa: A Dialogue on Development

The WCC through the PWE project, together with the Helsinki Process, the Chinese Academy of Social Sciences, the African Union, the All Africa Conference of Churches, the Economic Justice Network of FOCCISA and Beacon, convened in June 2011 the European Union-China-Africa Development Dialogue (ECADD) to address the concern over the new "repartitioning" of Africa for economic interests as manifested in the Economic Partnership Agreements between the European Union (EU) and Africa and China's growing appetite for Africa's raw materials and investments in oil, energy, mining and construction sectors in the region. The main question the dialogue seeks to address is: How does Africa attain economic liberation and secure socio-economic benefits as well as ecological protection in its partnerships, whether with the EU or with China? The ECADD aims to assess the involvement of the EU and China in African development with a view to measuring African benefit from EU and Chinese investments in Africa against the win-win principle.

From the European perspective, it was pointed out that European investments in Africa are relatively small – with South Africa, Nigeria and Egypt being the main investment destinations – compared to European investments in other regions. At the same time, European investments continue to be extremely important for African countries since asymmetric integration in the global economy leads to asymmetric negotiating power. Trade relations between the two regions were more significant: In 2009, 10 percent of all EU imports came from Africa.

From the Chinese perspective, there was acknowledgement of a growing international critique of China's involvement in Africa with the purported aim of securing the region's resources for raw materials. In fact, however, China has sourced from Africa a smaller amount (in both absolute and percentage terms) of energy and resources compared with the EU and the United States. Chinese companies hold under 2 percent of Africa's known oil reserves, with the vast majority being under the control of Western oil companies. Moreover, China is sensitive about African sovereignty and has a policy of non-interference in domestic concerns of African countries.

From the African perspective, it was felt that globalization has created a new scramble for Africa, with the colonists now in the form of foreign extractive and trading companies.

Presently the EU is negotiating EPAs with five regions in Africa: the Economic Community of West African States, Central African Economic and Monetary Community, Southern African Development

Community, Eastern and Southern African block, and East African Community (the EPA with the latter is already signed). Since the EU continues to be the major trading partner for many African countries, the challenges of liberalization for Africa have been substantial. African countries have to find a balance between reducing revenue losses and protecting domestic producers. In 2008, the EU launched the "Raw Materials Initiative," which aims to secure European access to cheap raw materials, and which is likely to deepen asymmetries in EU-Africa trade and investment relations.

On the other hand, dealing with China has some advantages, mainly the absence of "colonial baggage" and China's willingness to provide low-interest and "unrestricted" loans and aid. China's interest in Africa has given African nations more options to negotiate better trade deals with Western competitors. To some extent, Chinese presence has also permitted the building of local industries in some African states.

The downside of China's involvement in Africa has to do with the "occupation" by Chinese military or security personnel (especially in the Sudanese oil-producing areas) and the sale of weapons to some African governments that use these against their own civilians.

In moving forward, the ECADD is expected to address issues emerging from the discussions thus far. Among others, participants in the dialogue saw the need to deepen interrogation of the values that underpin development as practiced today by both the EU and China.

Encounters with the World Bank and International Monetary Fund

As part of the PWE process and building on previous encounters with the World Bank (WB) and the International Monetary Fund (IMF), the WCC hosted a dialogue in Accra in November 2008 between farmers and representatives of the WB and IMF on the issue of trade liberalization in Ghana as well as invited these international financial institutions to share their views on the global financial crisis with churches in Geneva in February 2009.

During the Ghana meeting in November 2008, which focused on the liberalization of the agricultural sector, local farmers challenged WB and IMF reports demonstrating how the opening up of agricultural markets in Ghana has led to progress. The farmers presented counter figures from the ground to show that free trade policies had, on the contrary, led to the violation of the right to food and livelihoods for people in the rural areas. Responding to the issues raised by the farmers, the WB pointed out that they now recognize the importance of government subsidies and other forms of support to the agricultural sector in some situations. At the end of the dialogue, participants expressed the need to see how these changes in WB trade policy would be actually implemented in Ghana and advocated for the organization of a national platform comprised of churches and civil society to challenge government, WB and IMF policies seeking to intensify trade liberalization.

At the discussion in Geneva in February 2009 on the theme of "Financial Crisis and the Way Out," the WB's Special Representative to the UN and the World Trade Organization and the Director of the IMF offices in Europe admitted that the market could not be left alone to solve global economic problems. Representatives from churches raised critical questions on the WB's and IMF's proposed solutions to the crisis, which still did not address the root causes of the crisis.

Engagement with the United Nations and Related Organizations on Financing for Development, Millennium Development Goals and Decent Work

The WCC through the PWE process has dialogued with governments and intergovernmental agencies through its engagement in various UN initiatives such as: the Financing for Development process, the initiative on the Millennium Development Goals (MDGs), and the International Labour Organization (ILO)'s campaign on Decent Work.

On the occasion of the 52nd session of the UN Commission on the Status of Women, which took place in New York in February 2008 under the theme of "Financing for Gender Equality and Women's Empowerment," a statement on financing for gender equality and development[4] was submitted on behalf of Ecumenical Women[5] which was received by the UN Secretary-General and member governments as an official document. The statement emphasized the need for just economic relationships between countries of the North and South and between women and men as well as lifted up alternative policies that prioritize provisioning for life in the six areas/themes outlined in the Monterrey Consensus. The statement served as the basis for drafting and advocating for Ecumenical Women's proposed language revisions to the "Agreed Conclusions and Recommendations" of the 52nd CSW.

A small ecumenical team composed of representatives from South Africa and the Philippines was convened to engage in the Follow-up Review Conference on Financing for Development held in Doha in November 2008. The ecumenical team prepared a statement entitled "The Time for Justice Is Now," which reiterated concerns raised by the WCC at the first Financing for Development conference held in Monterrey in 2002 and highlighted the failure of the neoliberal development paradigm using the global food, financial and climate crises as cases.[6] The statement:

> urges the UN to take leadership in redesigning an international financial architecture that establishes a global system of regulation as well as enlarges the space for developing country governments to enhance social protection in crisis periods... call[s] for a lasting solution to the debt problem for poor and middle-income countries beginning with the unconditional cancellation of the

---

[4] The statement to the 52nd CSW is available at http://daccess-dds-ny.un.org/doc/UNDOC/GEN/N07/659/16/PDF/N0765916.pdf?OpenElement

[5] Ecumenical Women is composed of the Anglican Communion, Church World Service, Episcopal Church (USA), the Lutheran World Federation, the National Council of Churches (USA), the Presbyterian Church (USA), United Methodist Church, United Church of Christ, World Council of Churches, World Federation of Methodist and Uniting Church Women, World Student Christian Federation, and the World YWCA.

[6] The statement entitled "The Time for Justice is Now" may be downloaded at http://www.oikoumene.org/en/resources/documents/wcc-programmes/public-witness-addressing-power-affirming-peace/poverty-wealth-and-ecology/finance-speculation-debt/statement-on-the-doha-outcome-document.html.

illegitimate debts being claimed from poor countries…as well as the setting up of an independent and fair debt arbitration mechanism, under UN auspices, for current and future loans to promote ethical lending and borrowing… call[s] for the removal of structural inequalities in the global trade system and the establishment of mutuality, transparency and civil society (not least women's) participation in future negotiations…[and] urges the UN to conceive a just and sustainable climate change financing framework that redistributes the financial outlays for mitigation and adaptation among and within countries in proportion to their contribution to climate change and according to their capacity to pay.

At the UN General Assembly Hearing with Civil Society on the MDGs in June 2010, a statement was presented, emphasizing the urgency of addressing the structural roots of the scandal of poverty and calling on governments to pursue economic policies as well as build economic frameworks that move away from the current paradigm that is focused on unlimited growth and based on structural greed toward models founded on pro-poor, redistributive growth; universal provisioning of common social goods; sustainable consumption and production; and investments in small-holder agriculture (which continues to be the main source of livelihood for people and women in poverty), social reproduction and ecological protection.[7]

Likewise, on the occasion of the UN Review Summit on the MDGs in September 2010, a letter addressed to the UN General Assembly from the WCC General Secretary was issued, proposing alternative approaches to meeting the MDGs by focusing on transforming unjust trade and financial policies and systems.[8]

---

[7] The statement on the MDGs may be accessed at
http://www.oikoumene.org/en/resources/documents/wcc-programmes/public-witness-addressing-power-affirming-peace/poverty-wealth-and-ecology/statement-on-the-millenium-development-goals.html.
[8] The letter to the UN Secretary-General is available at
http://www.oikoumene.org/en/resources/documents/general-secretary/messages-and-letters/letter-to-the-un-secretary-general-on-millenium-devolpment-goals.html.

Together with the Pontifical Council for Justice and Peace and the Islamic Educational Scientific and Cultural Organization, the WCC collaborated with the ILO in developing a policy handbook entitled *Convergences: Decent Work and Social Justice in Religious Traditions* (2012).[9] The dignity of work and workers is a common value among faith traditions. The handbook encourages policymakers to work with faith communities for social protection and security for all.

Engagement with Civil Society at the World Social Forum

Engaging with civil society and people's movements on issues of poverty, inequality and ecological degradation has been an important methodology employed in implementing the PWE project. The World Social Forum (WSF) is one of the primary venues where interaction, dialogue and cooperation with movements for justice and peace have taken place. Ecumenical presence at the WSF is geared toward building solidarity with civil society as well as showing the common witness of Christians in the current world climate – which does not claim to be "better" voice, but remains an authentic alternative voice. From the WSF in 2007 in Nairobi to the WSF in Dakar in 2011, seminars and campaigns have been co-organized with ecumenical and civil society partners on topics such as illegitimate and ecological debt, the food crisis and the financial crisis to promote discussion and joint actions (see Box 7).

---

[9] The handbook may be downloaded at http://www.ilo.org/wcmsp5/groups/public/@dgreports/@exrel/documents/publication/wcms_172371.pdf.

> 7. Working with civil society on just finance
>
> With more than 20 organizations involved in discussing and advocating for just finance, the WCC co-organized three seminars at the WSF 2009 held in Belem and co-issued a statement entitled "For a New Economic and Social Model: Let's Put Finance in Its Place"[10] which was signed by more than 600 organizations and sent to the G20.

---

[10] The statement entitled "Let's Put Finance in Its Place" may be downloaded at http://www.choike.org/campaigns/camp.php?5.

5.

Sharing Practical Approaches: Areas of Concern

Harvesting Good Practices from Churches and Ecumenical Partners

One of the aims of the PWE project is to lift up and harvest good and practical approaches from churches and ecumenical partners that address the areas emphasized in the AGAPE Call. The critical task ahead is to propagate, expand, and step up these practices as part of building a political movement for an Economy of Life. Though hardly an exhaustive compilation, the following are some inspiring and innovative examples.

Critiquing Neoliberal Globalization and Addressing Poverty Eradication

Responding to the phenomenon of economic globalization and the scandal of poverty, some churches around the world have delivered statements seeking for a just world and an Economy of Life. In 2007, the Church of Norway issued a resolution,[1] accompanied by a document entitled "The Church and Economic Globalization," which states:

> We confess that as a church in the North we are part of a tradition that makes us responsible for a political and economic development based on gross exploitation of human beings and natural resources in the South. We need to recognize more profoundly how this injustice has come about, and what we can do to create a more just world. We acknowledge that the present global trade and finance institutions… do not sufficiently protect the needs and interests of the poorest. These global institutions must be reformed and democratized…As a church in the North we want to be in solidarity with those who are suffering under

---

[1] The Church of Norway resolution and document on economic globalization is available at http://www.kirken.no/english/news.cfm?artid=162819.

> the injustice caused by the present global trade and finance system. We urge Norwegian authorities, business, organizations and other actors to have a holistic and concrete approach to the challenges of globalization...where reduction of poverty and environmental efforts are seen in relation to each other...[and] where efforts are made to reach a binding global human rights regime and democratic and legitimate supranational arrangements...

In 2009, the Uniting Church of Australia released a statement entitled "An Economy of Life: Reimagining Human Progress for a Flourishing World."[2] It states:

> The Church's role in the mission of God in the world requires us to be constantly asking how we might be good news in our own world. We are challenged to do this in a way that acknowledges both the huge problem of economic poverty within its complex causal context (historical, economic, social and environmental) and the impoverishment of all peoples, through the tyranny of greed, abuse, violence and obsession, disconnecting us from each other and the natural world. We are called to live out the vision of Jesus for human wholeness, an alternative understanding of what constitutes human progress – the love of God, made manifest for people in the experiences of dignity and respect, meaning and purpose in life, connection with the Earth and all its creatures, health and security of person and inclusion in communities of care and participation in society.

Churches in Europe jointly prepared and presented *proposals for addressing poverty* in the EU in 2010 in the context of the European Year for Combating Poverty and Social Exclusion (Pavlovic, ed. 2011). Key recommendations include the need for the European Commission to

---

[2] The Uniting Church of Australia Statement may be accessed at http://www.unitingjustice.org.au/just-and-sustainable-economy/uca-statements/item/461-an-economy-of-life-re-imagining-human-progress-for-a-flourishing-world.

develop a system of minimum wages and minimum income schemes and to invest more in the protection of poor people in the EU budget.

Churches are also interrogating poverty in creative ways. The Church of Scotland sponsored the Poverty Truth Commission,[3] a two-year project initiated in 2009 which brings together people living in poverty and civic leaders in order to discover the truths about poverty, and to explore real solutions to it. The Commission has focused on three areas deemed relevant to the Scottish context: overcoming violence, kinship care and stereotyping poverty.

At the same time, grassroots networks in partnership with churches are beginning to build alternative economic models that serve life. The Szécsény Social Cooperative in Hungary offers secure and sustainable livelihoods for its members while the *solidarity economy* movement, which originated in Brazil with the involvement of churches, works toward an economic system "centred on humankind and its longing to achieve full human rights (starting with the right to a decent life) and to meet individual, family, social and collective needs effectively on harmonious terms with nature and society" (Arruda 2008).[4] In Brazil, some 1.2 million workers are participating in the solidarity economy with activities ranging from agricultural production to crafts, apparel, microcredit cooperatives, bankrupt companies that have been salvaged by workers' cooperatives, community church projects and university incubators of solidarity businesses. In the last five years, more than a thousand enterprises based on solidarity economy principles have emerged in Brazil, but the network is growing internationally. In such enterprises, property and management rights are given by work, not by capital and production processes, and must be respectful of labour rights and ecological constraints. In the long run, the idea is to transform the model of production and consumption and alter direction of development toward a model that supports rather than is harmful to life.

---

[3] More information on the Poverty Truth Commission is available at http://www.povertytruthcommission.org/.
[4] See *Exchanging Visions of a Responsible, Plural, Solidarity Economy* by Marcos Arruda (2008).

Similarly, in the United States, grassroots movements within church communities are emerging around advocacy for specific public policies, eco-holism, and *intentional communities*. According to Elizabeth Hinson-Hasty (in Kennedy, ed. 2012, forthcoming):

> These three different grassroots approaches seek to envision alternatives to our current economic system by practicing communal decision making, grounding themselves in a consciousness of the radical interdependence of human beings with the whole creation, modeling a life lived for and with others, and sharing wealth and goods for any who might have need. Within grassroots movements such as these, *koinonia*, authentic community known in shared partnership, emerges from the perspective of people living at the vanishing point of today's economic picture. Moreover, eco-holism and intentional communities are creating change within the church itself as they bring Christian sources of thought and practice into conversation with other sources of wisdom (including wisdom from a variety of religious traditions and social organizations) and increase their mission, vision, and means through consensus-building models of decision making.

Addressing Trade

The Economic Justice Network of Fellowship of Councils and Churches in Southern Africa (FOCCISA) continues to *advocate for just trade* and *monitors and assesses Economic Partnerships in Africa (EPAs)* between the European Union and Africa, which demand deep cuts in tariffs from African nations as well as the opening up of African markets to subsidized EU produce. [5]

Many churches around the world are also engaged in *fair trade* (see Box 8).

---

[5] See http://www.ejn.org.za/index.php/our-projects/trade-and-justice for more details on EJN's work on EPAs.

> 8. Swedish churches for fair trade
>
> In 2005 there were not many people in the small Swedish town of Gunnarsbyn, north of Boden, who knew what fair trade was. Today, the shop is significant for the whole parish and people travel from all over the neighbouring counties...to shop. Goods from the South are neatly stacked on the shelves or hanging from hooks. Today, over 20 people work in the shop as volunteers. They give of their time to work in the shop and many of them have taken part in study groups about fair trade...Everything started with a 5000 kr grant from the local church council. With that money, goods were bought that fitted on one shelf in a corner of the parish building...Now the Worldshop in Gunnarsbyn takes up two rooms...with over 1000 products in stock...In 2009 Gunnarsbyn parish was amongst the first of five parishes to receive the award...for Fair Trade Church.
>
> - Excerpted from "Justice in Practice –
> The Church of Sweden's Work on Sustainable Lifestyles"
> (Oreland et al. in Pavlovic, ed. 2011)

Addressing Finance

Churches in Latin America are tireless in their *advocacy for the cancellation of illegitimate debt* as well as for the development of an international independent court for the just arbitration of external debt. The CLAI was actively involved in Ecuador's Integral Audit of the Public Debt in 2008. On the basis of the audit, the Ecuadorian government refused to pay some of the most unjust debts being claimed from them. In solidarity, churches in Norway and Germany have challenged their governments on concrete cases of illegitimate debt, resulting in the Norwegian government's write off of parts of debts being claimed from Ecuador, Peru and Jamaica (Pavlovic, ed. 2011).

Since the onset of the global financial crisis in 2008, more churches are *supporting and developing campaigns for the implementation of a financial transaction tax at national, regional and global levels* in partnership with civil society. In June 2012, the LWF issued a resolution encouraging member churches to advocate with their governments for the creation of such a tax which would not only help to curb speculation, but could also raise public funds for poverty eradication and ecological protection.[6] The governments of Austria, Belgium, Italy, Finland, France, Germany, Greece and Portugal appear to be supportive of such a tax.[7]

Churches are also *advocating for the use of alternative indicators to measure socio-economic well-being other than monetary-based indicators*. The Federation of Swiss Protestant Churches is *addressing the issue of the exceedingly high compensation and bonuses received by top-level managers and directors in the corporate and banking sector* through the issuance of voluntary guidelines. Moreover, churches are increasingly taking part in *ethical investments*. For instance, the Church of Sweden is currently looking at the climate impact of their shareholdings with a view to increasing investments in companies which work toward sustainable energy solutions (Oreland et al in Pavlovic, ed. 2011).

Addressing the Sustainable Use of Land and Natural Resources

Churches in Africa – Mozambique, Tanzania, South Africa, and Zambia, among others – have been actively engaged in confronting government policy and multinational mining corporations that seek to exploit the region's rich resources with minimal contributions to government revenues and to national employment and at huge price to the environment. Tanzanian churches have spoken up against uranium mining.[8] Since 2010, the Economic Justice Network of FOCCISA with the support of Norwegian churches has been convening an annual Alternative Mining Indaba in Cape Town that brings together people and organizations from mining communities in Africa and around the

---

[6] See http://blogs.lutheranworld.org/wordpress/rioplus20/council-2012-resolution-on-financial-transactions-tax/ for the LWF Council resolution in support of financial transaction tax. The Uniting Church of Australia also backs the tax (see http://assembly.uca.org.au/news/item/674-uniting-church-supports-robin-hood-tax).

[7] See http://www.spiegel.de/international/europe/fears-of-competitive-disadvantage-euro-zone-split-over-financial-transaction-tax-a-820965.html.

[8] For details, see http://www.uranium-network.org/index.php/archive-2012/194-tanzanian-churches-oppose-uranium-mining.

world in counterpart to the annual Mining Indaba that is attended by mining investors.[9] The Alternative Mining Indaba is a space for theological reflection on resource extraction as well as for reviewing mining policies from the perspective of impacts on communities and ecology and discussing alternatives. The 2013 Alternative Mining Indaba could be a space to present ecological debt cases in the mining sector with a view to building a *campaign to demand for holistic reparations* to address ecological debt. Spearheaded by churches, national versions of the Alternative Mining Indaba have taken place in Zambia and Mozambique.

Similarly, churches in Asia are scrutinizing large-scale mining operations in their countries, while churches in Canada are taking Canadian mining companies to task for their unsustainable operations in other countries. In March 2012, Philippine churches including Roman Catholic and Protestant churches convened the Eastern Visayas Ecumenical Forum on People's Mining resulting in a statement calling for the repeal of the current mining law which is deemed to be anti-people and anti-ecology and to educate and mobilize parishioners in defense of their communities and God's Creation.[10] Canadian churches also organized an Ecumenical Conference on Mining in April 2011, culminating in a statement which observed that: "As churches, we recognize our internal contradictions and complicity with respect to resource extraction, and the urgent need to practice responsible consumption and citizenship."[11]

Churches in India continue to voice their *opposition to the multi-billion dollar POSCO steel project* in Orissa that is projected to displace at least 22,000 farmers, adversely affect the livelihoods of 20,000 fisher folk, generate a water crisis in the area, harm already threatened animal species, and deplete forests where *adivasis* or tribal people dwell and derive

---

[9] For more information on the Alternative Mining Indaba, please visit http://www.ejn.org.za/index.php/component/content/article/61-current-feature-cluster/569-just-miningor-just-profits.
[10] See http://www.samarnews.com/news2012/mar/b706.htm for details on the Philippine ecumenical conference on mining.
[11] For additional information on the Canadian ecumenical mining conference, visit http://www.kairoscanada.org/sustainability/resource-extraction/ecumenical-conference-on-mining/.

sustenance from. The struggle against POSCO has been taken up by Oikotree,[12] a growing network of faith-based organizations and people's movements borne out of the AGAPE Call and the Accra Confession and initiated by the WCC, WCRC and CWM that aims to put justice at the heart of faith. Under the Oikotree umbrella, Korean churches are getting involved in the campaign against POSCO, which is a Korean company.

The World Student Christian Federation (WSCF) is developing a global campaign for eco-justice. The Latin America and Caribbean and the North American regions of the WSCF held a joint programme called "Students on an Encounter for Solidarity and Eco-justice" in Cuba in April 2012, which discussed alternative paradigms such as the indigenous concept of *buen vivir* (good living) and eco-feminism as well as proposed strategies to address growing state and transnational exploitation of nature.[13] Moreover, the WSCF and the Agape Centro Ecumenico, an initiative of the Waldensian Churches in Italy, organized *the Agape International Political Camp with a focus on eco-justice* in Prali in August 2012. The camp studied "the current ecocide" of the planet by considering patterns of production and consumption, in particular heeding the call of Indigenous Peoples that have never stopped trying to live in harmony with the planet," reflected on the intrinsic link between economy and ecology and the roles of science and technology, Indigenous Peoples and religions in addressing the crisis, as well as discussed sustainable and regenerative alternatives to the current order and how to implement them.[14]

In Germany, churches, trade unions, environmental and other groups are beginning to come together to discuss in their communities the concept of *"degrowth" or "zero growth"* as a way to stabilize levels of material consumption and production in a planet characterized by finite resources as well as a finite capacity for absorbing waste. Dialogue is

---

[12] Visit www.oikotree.org for more information on the network.
[13] See http://wscfna.org/news/north-american-students-eco-justice-cuba for details on the outcomes of this activity.
[14] More information on the Agape International Political Camp on eco-justice can be accessed at http://agapecentroecumenico.org/en/what-is-it-we-do/agapes-camps/international-political-camp-2012/.

crucial as such a radical approach could have adverse effects on certain sectors and will entail sacrifices. In the long run, however, a "degrowth" strategy in rich, industrialized nations could be pivotal to developing an "economy of enough."

Churches in Latin America and elsewhere are also starting conversations on *Earth rights*, in counterpart to human rights, based largely on the indigenous Cosmo vision that sees humans as part of creation. In 2011, Bolivia became the first nation to give all nature equal rights to humans under the Law of Mother Earth. Ecuador recently made changes to its constitution to grant nature "the right to exist, persist, maintain and regenerate its vital cycles, structure, functions and its processes in evolution." However, many challenges remain in terms of implementing and realizing Earth rights.

Public Goods and Services

Churches in Latin America are *resisting the privatization of public goods*, particularly water. At the Ecumenical Water Forum held in Belem in 2009, the National Council of Christian Churches of Brazil affirmed the responsibility of churches to protect water as a public good and a human right.[15]

Likewise, churches in the United States have contributed to the debate on health care reform in the country. In particular, the United Methodist Church affirmed *healthcare as a public good and human right*: "We…recognize the role of governments in ensuring the each individual has access to those elements necessary to good health."[16]

Life-Giving Agriculture

Many churches both in the North and the South are engaged in *advocacy on food sovereignty and on food as a human right* in various forums from the

---

[15] For details on Brazilian churches' position on water privatization, see http://www.ekklesia.co.uk/node/8534.

[16] See http://www.umc-gbcs.org/site/apps/nl/content3.asp?c=frLJK2PKLqF&b=3631781&ct=3956183 for the United Methodist Church's position on the health care debate in the United States.

WSF to the UN and through the Ecumenical Advocacy Alliance which has "food for life" as one of its focal points from 2009 to 2012. In particular the Fellowship of Councils of Churches in West Africa (FECCIWA) has launched a campaign called "Eat what you grow, grow what you eat" to realize the basic human right to be free from hunger.[17] Increasingly these campaigns are linked to campaigns addressing climate change, land grabbing and the proliferation of plantations for biofuels.

Churches in Korea continue to spearhead the *life-giving agriculture movement*, supporting farmer's movements, promoting the cultivation of organic, locally grown produce and establishing rural cooperatives. The movement does not only endeavour to make changes in dominant agricultural technologies based on chemical fertilizers and pesticides that are destructive to human health and ecology, but also aims to transform culture, recognizing that "without agriculture, there is no culture."

Churches are also *advocates for land reform*. The Church of North India collaborated with several other civil society organizations in sponsoring "Janedesh 2007" – a 26-day march of 25,000 landless persons from 15 states to the capital, Delhi, to demand for the redistribution of land toward the tillers, the formulation of a national land policy and a speedy and just resolution of land disputes.[18]

Decent Jobs and Livelihoods

Churches and related organizations are in the forefront of *campaigns for migrant workers' rights and decent jobs and livelihoods*. In Hong Kong, India, Indonesia, Korea, Philippines, Singapore, Thailand, Canada and elsewhere, churches are advancing the rights of migrant workers and are fighting against human trafficking. In particular, the Centre for Indonesian Migrant Workers of the Urban Community Mission of Jakarta is involved in training and re-integrating migrant workers and

---

[17] See http://fecciwa.org/index.php?option=com_content&view=article&id=61%3Afood-security&catid=36%3Acampaings&Itemid=80&lang=en for more information on the FECCIWA campaign.

[18] See http://globalministries.org/news/sasia/church-of-north-india-marching.html for more information on "Janedesh 2007."

their families, running a crisis centre, conducting research, building networks, organizing communities and advocating for migrants' rights.

In 2011, member churches of the WCRC joined a *peaceful protest in solidarity with agricultural workers in Florida, called the Coalition of Immokalee Workers*, seeking a penny-per-pound increase on the price of tomatoes they grow.[19] Also in 2011, churches in the Philippines convened a *Church People-Workers' Solidarity* conference.[20] Philippine churches are standing on the side of workers of the Swiss food giant, Nestlé, which, to this date, has refused to comply with a 2008 Supreme Court decision to allow a decent retirement plan to be included in the collective bargaining agreement for factory workers in the Philippines.

Empire

The United Church of Canada released the report entitled "Living faithfully in the midst of empire" at the end of 2006.[21] The report describes the complexity of economic globalization and how such processes are exponentially increasing the pain and misery experienced by the vast majority of God's people. Following the issuance of the report, the United Church of Canada embarked on three years of engagement and animation as well as produced a number of educational materials on empire.

---

[19] See http://www.wcrc.ch/node/564 for more details on the protest in support of Immokalee workers.
[20] For details, visit http://hronlineph.com/2011/09/18/statement-church-people-and-workers-in-solidarity-reclaiming-the-dignity-of-human-work/.
[21] The report may be downloaded at http://www.united-church.ca/files/economic/globalization/report.pdf.

6.

God of Life, Lead Us to Justice and Peace: A Conclusion

In the run-up to and beyond the 10th Assembly of the WCC set to take place in Busan under the theme "God of life, lead us to justice and peace, " the work on linking poverty, wealth and ecology and promoting justice in the economy, society and in the Earth has never been more relevant and crucial. Such efforts are, at core, a matter of faith.

The global financial and economic meltdown in 2008 that continues to generate tremendous hunger and suffering and the looming climate crisis that poses severe threats to human and planetary survival have borne out the WCC's prophetic critique of unfettered market liberalization and deregulation, the obsession with economic growth, the pursuit of profit at all cost and the culture of greed, consumerism and accumulation that are the hallmarks of the prevailing economic paradigm.

But the work of transformation does not end with critique. Churches have a vital role to play in fostering the moral courage essential for witnessing to a spirituality of justice and care for all creation and building synergies for a political movement advocating for an Economy of Life. The "seeds of hope" are already being planted as evidenced by the wide range of church efforts to address interconnected problems of poverty, inequality and ecological destruction. Notwithstanding this, it is clear that plenty more work remains to be done in order to develop far-reaching, more cohesive and coordinated programmes geared at changing economic structures, production and consumption patterns, cultures and values in the service of life.

The 2012 AGAPE Call to Action for "Economy of Life, Justice and Peace for All" declares:

God calls us to a radical transformation. Transformation will not be without sacrifice and risk, but our faith in Christ demands that we commit ourselves to be transformative churches and transformative congregations…The process of transformation must uphold human rights, human dignity and human accountability to all of God's creation. We have a responsibility that lies beyond our individual selves and national interests to create sustainable structures that will allow future generations to have enough. Transformation must embrace those who suffer the most from systemic marginalization…Nothing determined without them is for them. We must challenge ourselves and overcome structures and cultures of domination and self-destruction that are rending the social and ecological fabric of life. Transformation must be guided by the mission to heal and renew the whole creation.

Bibliography

Arruda, Marcos (2008), *Exchanging Visions of a Responsible, Plural, Solidarity Economy*, ALOE, RPSE and PACS: Rio de Janeiro

Brubaker, Pamela and Rogate Mshana (2010), *Justice Not Greed*, WCC: Geneva

Church of Norway (2007), "The Church and Economic Globalization," retrieved from http://www.kirken.no/english/news.cfm?artid=162819

De Lange, Harry and Bob Goudzwaard (1995), *Beyond Poverty and Affluence: Toward an Economy of Care,* WCC: Geneva

ILO (2012), *Convergences: Decent Work and Social Justice in Religious Traditions*, ILO: Geneva

Kennedy, Joy, ed. (2012), *Poverty, Wealth and Ecology: Ecumenical Perspectives from North America,* WCC: Geneva.

Larrea, Carlos (2011), "Inequality, Sustainability and the Greed Line: A Conceptual and Empirical Approach," *Ecumenical Review* Vol. 63, Issue 3

LWF (2011), "Muslims and Christians Engaging Structural Greed Today: Conference Findings," September 2011, Kota Kinabalu, retrieved from http://www.lutheranworld.org/lwf/wp-content/uploads/2011/10/DTS-KotaKinabalu2011_FinalDoc.pdf

Mshana, Rogate (2007), *Poverty, Wealth and Ecology: The Impact of Economic Globalization*, WCC: Geneva, retrieved from http://www.oikoumene.org/fileadmin/files/wcc-main/documents/p3/poverty_24p.pdf

Mshana, Rogate(2008), *Wealth, Poverty and Ecology and their Links*, WCC: Geneva

Mshana, Rogate, ed. (2009), *Poverty, Wealth and Ecology: Ecumenical Perspectives from Latin America and The Caribbean*, WCC: Geneva

Mshana, Rogate, ed. (2012), *Poverty, Wealth and Ecology in Africa: Ecumenical Perspectives,* WCC: Geneva

Mshana Rogate, ed. (2013), *The European Union and China in Africa: The Development Dialogue*, WCC: Geneva

Pavlovic, Peter, ed. (2011), *Poverty, Wealth and Ecology in Europe: Call for Climate Justice*, Church and Society Commission of the Conference of European churches: Brussels

Peralta, Athena, ed. (2006), *Ecological Debt: The Peoples of the South Are the Creditors: Cases from Ecuador, Mozambique, Brazil and India*, WCC: Quezon City.

Peralta, Athena, ed. (2010), *Poverty, Wealth and Ecology in Asia and the Pacific: Ecumenical Perspectives*, WCC, CCA and PCC: Geneva, Chiang Mai and Suva, retrieved from http://www.cca.org.hk/resource/books/olbooks/poverty_wealth_and_ecology_in_ap.pdf

Pushra, Werner and Sarah Burke, ed. (2011), *New Directions for International Financial and Monetary Policy: Reducing Inequalities for Shared Societies*, FES: New York

Raiser, Konrad (2011), "Theological and Ethical Considerations regarding Wealth and the Call for Establishing a Greed Line," *Ecumenical Review* Vol. 63, Issue 3

Sung, Jung Mo (2011), "Greed, Desire, and Theology," *Ecumenical Review* Vol. 63, Issue 3

Taylor, Michael, ed. (2003), *Christianity, Poverty and Wealth*, WCC: Geneva

The Ecumenical Review: Greed and Global Economics, 63:3 (October 2011)

Ul Haque 2004, "Globalization, Neoliberalism, and Labour," UNCTAD Discussion Paper, retrieved from http://www.eviangroup.org/m/1392.pdf

United Church of Canada (2006), "Living Faithfully in the Midst of Empire," retrieved from http://www.united-church.ca/files/economic/globalization/report.pdf

Uniting Church of Australia (2009), "An Economy of Life: Reimaging Human Progress for a Flourishing World," retrieved from http://www.unitingjustice.org.au/just-and-sustainable-economy/uca-statements/item/461-an-economy-of-life-re-imagining-human-progress-for-a-flourishing-world

WARC, now WCRC (2004), "The Accra Confession – Covenanting for Justice in the Economy and the Earth," retrieved from http://www.warc.ch/documents/ACCRA_Pamphlet.pdf

WCC (2006), *Alternative Globalization Addressing Peoples and Earth (AGAPE): A Background Document.* rev. ed, Geneva

WCC (2008), "The Time for Justice Is Now," statement on the occasion of the UN Follow-up Review Conference on Financing for Development, December 2008, Doha, retrieved from http://www.oikoumene.org/en/resources/documents/wcc-programmes/public-witness-addressing-power-affirming-peace/poverty-wealth-and-ecology/finance-speculation-debt/statement-on-the-doha-outcome-document.html

WCC (2008), "Financing for Gender Equality and Development," statement on the occasion of the 52nd Commission on the Status of Women, February 2008, New York, retrieved from http://daccessdds.un.org/doc/UNDOC/GEN/N07/659/16/PDF/N0765916.pdf?OpenElement

WCC (2009), "Eco-justice and Ecological Debt," central committee statement issued in September 2009, Geneva, retrieved from http://www.oikoumene.org/en/resources/documents/central-committee/geneva-2009/reports-and-documents/report-on-public-issues/statement-on-eco-justice-and-ecological-debt.html

WCC (2009), "Just Finance and an Economy of Life," central committee statement issued in September 2009, Geneva, retrieved from http://www.oikoumene.org/en/resources/documents/central-committee/geneva-2009/reports-and-documents/report-on-public-issues/statement-on-just-finance-and-the-economy-of-life.html

WCC (2009), "Letter to G20 Countries on the Global Economic and Financial Crisis," April 2009, Geneva, retrieved from http://www.oikoumene.org/en/resources/documents/general-secretary/messages-and-letters/27-03-09-letter-to-g20.html

WCC (2010), "The WCC's Statement on Occasion of the UN General Assembly Hearing with Civil Society on the MDGs," June 2010, New York, retrieved from http://www.oikoumene.org/en/resources/documents/wcc-programmes/public-witness-addressing-power-affirming-peace/poverty-wealth-and-ecology/statement-on-the-millenium-development-goals.html

WCC (2011), "Glory to God and Peace on Earth – Message of the International Ecumenical Peace Convocation," May 2011, Kingston, retrieved from http://www.overcomingviolence.org/en/resources-dov/wcc-resources/documents/presentations-speeches-messages/iepc-message.html

WCC (2011), "Social Justice and Common Goods," CCIA policy paper, retrieved from http://www.oikoumene.org/resources/documents/wcc-commissions/international-affairs/economic-justice/social-justice-and-common-goods-policy-paper.html

WCC (2012), "Economy of Life, Justice and Peace for All: A Call to Action," July 2012, Bogor, retrieved from http://www.oikoumene.org/en/resources/documents/wcc-programmes/public-witness-addressing-power-affirming-peace/poverty-wealth-and-ecology/neoliberal-paradigm/agape-call-for-action-2012.html

WCC (2012), "Statement on the Current Financial and Economic Crisis with a Focus on Greece," central committee statement issued in September 2012, Crete, retrieved from http://www.oikoumene.org/gr/resources/documents/central-committee/kolympari-2012/report-on-public-issues/viii-statement-on-the-financial-crisis.html

WCC and LWF (2010), "Buddhist Christian Common Word on Structural Greed," August 2010, Chiang Mai, retrieved from http://www.oikoumene.org/en/resources/documents/wcc-programmes/interreligious-dialogue-and-cooperation/interreligious-trust-and-respect/buddhist-christian-common-word-on-structural-greed.html

WCC, WCRC and CWM (2012), "The Sao Paolo Statement: International Financial Transformation for the Economy of Life," October 2012, Sao Paolo, retrieved from http://www.oikoumene.org/en/resources/documents/wcc-programmes/public-witness-addressing-power-affirming-peace/poverty-wealth-and-ecology/finance-speculation-debt/sao-paulo-statement-international-financial-transformation-for-the-economy-of-life.html

Appendix 1

PWE Consultations 2007-2012

Consultation on Poverty, Wealth and Ecology in Africa, 05-09 November 2007, Dar es Salaam

Organized in cooperation with the All Africa Conference of Churches and Christian Council of Tanzania from 05-09 November 2007 in Dar es Salaam, the consultation on poverty, wealth and ecology in Africa was attended by 65 church representatives predominantly from the continent of Africa. Participants from Asia, Latin America, Europe and North America shared experiences from their countries and regions. The Dar es Salaam gathering reflected on a theological base for studying the links between poverty, wealth and ecology in Africa; discussed the perspectives of African women and youth on the links between poverty, wealth and ecology; shared standpoints on poverty, wealth and ecology of an African state leader, an African woman in poverty and a wealthy African man; considered the findings of study entitled "Wealth Creation, Poverty and Ecology in Africa;" exchanged North-South and South-South Church perspectives and practices on the causes of and ways of tackling poverty, inequality, and environmental destruction; and came up with joint church strategies and actions for addressing the interlinked problems of poverty, excessive wealth, and ecological degradation in the region. The Dar es Salaam Statement calls on "those who collude with systems of domination in economy and ecology - including African government leaders and elites - to recognize, confess, repent and engage in restorative, distributive and transformative justice."

Consultation on Poverty, Wealth and Ecology in Latin America and the Caribbean, 06-10 October 2008, Guatemala City

The meeting in Guatemala City gathered more than 50 participants from churches (including the Roman Catholic Church) in the Latin America and Caribbean region accompanied by a few representatives from Africa,

Asia-Pacific, Europe and North America. Two-day long hearings of youth, women and Indigenous Peoples took place to discuss the perspectives of these groups on poverty, wealth and ecology. One of the keynote presentations was given by Rosalinda Tuyuc, an indigenous leader of the women's organization CONAVIGUA or the Guatemalan Council of Widows, who pointed out that according to the Mayan Cosmo vision, all life is interlinked such that the destruction of the forests in pursuit of development will have adverse implications on people's survival. The meeting produced the Guatemala Statement which acknowledged that the consultation was taking place in a period characterized by global financial, food and climate crises, which also represents a crisis of the neoliberal model. At the same time, it highlighted the gradual retreat of neoliberalism in the region with the election of progressive leaders and increased intra-regional cooperation and solidarity.

Consultation on Poverty, Wealth and Ecology in Asia and the Pacific, 02-06 November 2009, Chiang Mai

Hosted by the Church of Christ in Thailand and jointly organized by the WCC, the Christian Conference of Asia and the Pacific Conference of Churches, the consultation on poverty, wealth and ecology in Asia-Pacific was held in Chiang Mai from 02-06 November 2009 and was attended by 80 church representatives drawn mostly from the women, youth and Indigenous Peoples hearings and accompanied by participants from Africa, Europe, Latin America and the Caribbean and North America. The Indigenous Peoples' hearing raised issues around dispossession of land and the impact of development on ecology. The women's hearing lifted up issues of migration, human trafficking and the gendered impacts of climate change. The youth hearing emphasized growing unemployment and heightened consumerism especially among young people. The Chiang Mai gathering included an enriching discussion on Buddhist, Hindu and Islamic perspectives on poverty, wealth and ecology. The consultation's final output, the Chiang Mai Declaration, is a distillation of the intensive discussions that took place and encapsulates recommendations for church advocacy and action. It calls on churches to shape Economies of Life and Economies for Life.

Consultation on Poverty, Wealth and Ecology in Europe, 08-12 November 2010, Budapest

Around 80 participants from churches and ecumenical partners participated in the Consultation on Poverty, Wealth and Ecology in Europe, which was held in Budapest from 08 to 12 November 2010. The consultation was organized in partnership with the Conference of European Churches (CEC) and hosted by the Ecumenical Council of Churches in Hungary. It highlighted an open and critical dialogue between the CEC and the CLAI on the threats and challenges of globalization as well as organized hearings around three themes, namely: (1) wealth and poverty in Europe, (2) facing up to a low carbon economy and an economy of sufficiency, and (3) dialogue with power structures and among churches: churches addressing the economic and financial crisis. The key document coming out of the consultation was the "Budapest Call for Climate Justice: Addressing Poverty, Wealth and Ecology," which states that churches must "put climate justice and poverty eradication and the relationship between the two as a priority on the agenda of its 10th General Assembly in South Korea in 2013." Additionally, it calls on governments in Europe to make climate justice a central goal of policy-making and to no longer see economic growth as an aim in itself. Aside from making more ambitious commitments to cutting GHG emissions, this would entail: "the redistribution of wealth and sharing of technology between rich countries and poor countries … with additional support for climate change mitigation and adaptation."

Consultation on Poverty, Wealth and Ecology in North America, 07-11 November 2011, Calgary

Taking place in Calgary, Alberta, Canada from 07 to 11 November 2011, the North America Consultation on Poverty, Wealth and Ecology – the last in a series of regional consultations – was preceded by a visit to the Alberta tar sands by a team of church representatives from all over the world. The visit helped many of the participants to critically analyze the intrinsic links between poverty, wealth and ecology in the North American context. The gathering issued a call to reflection and action in a time of global financial crisis, environmental threat, and resistance to the ways of Wall Street and its allied economic structures as reflected in the growing "Occupy" movement. In the final communique entitled

"There's a New World in the Making," the more than 80 representatives of North American churches urged the WCC and ecumenical partners "to undertake a decade of action on eco-justice encompassing both ecological and economic justice," recognizing that: "The cry of the poor and the cry of the Earth are one."

Global Forum on Poverty, Wealth and Ecology, 18-22 June 2012, Bogor

More than 100 participants from around the world attended the Global Forum on Poverty, Wealth and Ecology which took place in Bogor, Indonesia from 18 to 22 June 2012. The forum was hosted by WCC member churches in Indonesia, including Huria Kristen Batak Protestant (HKBP), Communion of Churches in Indonesia (PGI), Urban Community Mission Jakarta (PMK-HKBP) and Indonesia Christian Church (GKI-West Java regional synod). The Global Forum on Poverty, Wealth and Ecology came up with a call for action to evolve "transformative churches and transformative congregations" with the moral courage "to witness to a spirituality of justice and sustainability and build a prophetic movement for an economy of life." The call to action entitled "Economy of Life, Justice and Peace for All" is a follow-up to the AGAPE process. It captures key findings and affirmations from studies and consultations analyzing the interlinkages between poverty, wealth and ecology, undertaken in Africa in 2007, Latin America and the Caribbean in 2008, Asia and the Pacific in 2009, Europe in 2010 and North America in 2011.

Appendix 2

Statement on Eco-Justice and Ecological Debt

"Forgive us our debts, as we also have forgiven our debtors" (Matthew 6:12)

1. The era of "unlimited consumption" has reached its limits. The era of unlimited profit and compensation for the few must also come to an end. Based on a series of ecumenical consultations and incorporating the perspectives of many churches, this statement proposes the recognition and application of a concept that expresses a deep moral obligation to promote ecological justice by addressing our debts to peoples most affected by ecological destruction and to the earth itself. It begins with expressing gratitude to God, whose providential care is manifested in all God's creation and the renewal of the earth for all species. Ecological debt includes hard economic calculations as well as incalculable biblical, spiritual, cultural and social dimensions of indebtedness.

2. The earth and all of its inhabitants are currently facing an unprecedented ecological crisis, bringing us to the brink of mass suffering and destruction for many. The crisis is human-induced, caused especially by the agro-industrial-economic complex and culture of the global North, which is characterized by the consumerist lifestyles of the elites of the developed and developing worlds and the view that development is commensurate with exploitation of the earth's "natural resources." What is being labeled and co-modified, as "natural resources" is all of creation – a sacred reality that ought not to be co-modified. Yet the Northern agro-industrial-economic complex, especially in the current era of market globalization, has used human labour and resourcefulness, as well as the properties of other life forms, to produce wealth and comfort for a few at the expense of the survival of others and their dignity.

3. Churches have been complicit in this history through their own consumption patterns and through perpetuating a theology of human rule over the earth. The Christian perspective that has valued

humanity over the rest of creation has served to justify the exploitation of parts of the earth community. Yet, human existence is utterly dependent on a healthy functioning earth system. Humanity cannot manage creation. Humanity can only manage their own behaviour to keep it within the bounds of earth's sustenance. Both the human population and the human economy cannot grow much more without irreversibly endangering the survival of other life forms. Such a radical view calls for a theology of humility and a commitment on the part of the churches to learn from environmental ethics and faith traditions that have a deeper sense of an inclusive community.

4. The churches' strength lies in its prophetic witness to proclaim God's love for the whole world and to denounce the philosophy of domination that threatens the manifestation of God's love. The biblical prophets had long ago deduced the intrinsic connection between ecological crises and socio-economic injustice, railing against the elites of their day for the exploitation of peoples and the destruction of ecosystems (Jeremiah 14: 2-7, Isaiah 23: 1-24 and Revelations 22). Based on Jesus' commandment of love, as expressed in his life and parables, the World Council of Churches (WCC) must broaden its understanding of justice and the boundaries of who our neighbours are. For many years, the WCC has called for the cancellation of illegitimate external financial debts claimed from countries of the South based on the biblical notion of jubilee (Leviticus 23). It has taken a step further in addressing the ecological dimension of economic relationships.

5. Beginning with the articulation of the ideas of "limits to growth" in a Church and Society consultation held in Bucharest in 1974 and "sustainable societies" at the 1975 Nairobi assembly, the WCC has been working deeply on ecological justice for over three decades. At the 1998 Harare assembly, the harmful impacts of globalization on people and the environment came to the fore through the Alternative Globalization Addressing People and earth (AGAPE) process, leading to the on-going study process on Poverty, Wealth and Ecology. As an offshoot of these important ecumenical reflections and actions, the WCC, in partnership with churches and

civil society organizations in Southern Africa, India, Ecuador, Canada and Sweden, initiated work on ecological debt in 2002.

6. Ecological debt refers to damage caused over time to ecosystems, places and peoples through production and consumption patterns; and the exploitation of ecosystems at the expense of the equitable rights of other countries, communities or individuals. It is primarily the debt owed by industrialized countries in the North to countries of the South on account of historical and current resource plundering, environmental degradation and the disproportionate appropriation of ecological space to dump greenhouse gases (GHGs) and toxic wastes. It is also the debt owed by economically and politically powerful national elites to marginalized citizens; the debt owed by current generations of humanity to future generations; and, on a more cosmic scale, the debt owed by humankind to other life forms and the planet. It includes social damages such as the disintegration of indigenous and other communities.

7. Grounded on an overriding priority for the impoverished and a deep moral responsibility to rectify injustices, ecological debt lenses reveal that it is the global South who is the principal ecological creditor while the global North is the principal ecological debtor. The ecological debt of the global North arises from various causal mechanisms whose impact has been intensified in the current economic crisis.

8. Under the current international financial architecture, countries of the South are pressured through conditions for loans as well as multilateral and bilateral trade and investment agreements to pursue export-oriented and resource-intensive growth strategies. Ultimately it fails to account for the costs of erosion of ecosystems and increasing pollution. Many mega-development projects (e.g. dams) in countries of the South are financed through foreign lending by international financial institutions in collaboration with undemocratic and corrupt local leaders and elites, without the informed consent of local inhabitants and with little consideration of the projects' ecological and social consequences. Moreover, industrialized Northern countries make disproportionate use of ecological space

without adequate compensation, reparation or restitution. Northern countries' ecological footprint (an approximate measurement of human impacts on the environment) presently averages 6.4 ha/person. This is more than six times heavier than the footprint of Southern countries at an average of 0.8 ha/person.

9. Human-induced climate change heightens the relationship of North-South inequity even further. Industrialized countries are mainly responsible for GHG emissions causing climate change (though emerging economies in the South are becoming major contributors to global GHG emissions in absolute terms). Yet, research indicates that the South will bear a bigger burden of the adverse effects of climate change including the displacement of people living in low-lying coastal areas and small island states; the loss of sources of livelihood, food insecurity, reduced access to water and forced migration.

10. In the light of biblical teaching (cf. Matthew 6:12), we pray for repentance and forgiveness, but we also call for the recognition, repayment and restitution of ecological debt in various ways, including non-market ways of compensation and reparation, that go beyond the market's limited ability to measure and distribute.

11. The central committee of the WCC recognizes the need for a drastic transformation at all levels in life and society in order to end the ecological indebtedness and restoring right relationships between peoples and between people and the earth. This warrants a re-ordering of economic paradigms from consumerist, exploitive models to models that are respectful of localized economies, indigenous cultures and spiritualities, the earth's reproductive limits, as well as the right of other life forms to blossom. And this begins with the recognition of ecological debt.

While affirming the role of churches to play a critical role in lifting up alternative practices, as well as building the necessary political will and moral courage to effect urgent transformations, the central committee of the WCC meeting in Geneva, Switzerland, 26 August - 2 September 2009:

A. *Calls* upon WCC member churches to urge Northern governments, institutions and corporations to take initiatives to drastically reduce their greenhouse gas (GHG) emissions within and beyond the United Nations Framework Convention on Climate Change (UNFCCC), which stipulates the principles of historical responsibility and "common, but differentiated responsibilities" (CDR), according to the fixed timelines set out by the UNFCCC report of 2007.

B. *Urges* WCC member churches to call their governments to adopt a fair and binding deal, in order to bring the CO_2 levels down to less than 350 parts per million (ppm), at the Conference of Parties (COP 15) of the UNFCCC in Copenhagen in December 2009, based on climate justice principles, which include effective support to vulnerable communities to adapt to the consequences of climate change through adaptation funds and technology transfer.

C. *Calls upon* the international community to ensure the transfer of financial resources to countries of the South to keep petroleum in the ground in fragile environments and preserve other natural resources as well as to pay for the costs of climate change mitigation and adaptation based on tools such as the Greenhouse Development Rights (GDR) Framework.

D. *Demands* the cancellation of the illegitimate financial debts of Southern countries, most urgently for the poorest nations, as part of social and ecological compensations, not as official development assistance.

E. *Recommends* that WCC member churches learn from the leadership of Indigenous Peoples, women, peasant and forest communities who point to alternative ways of thinking and living within creation, especially as these societies often emphasize the value of relationships, of caring and sharing, as well as practice traditional, ecologically respectful forms of production and consumption.

F. *Encourages* and supports WCC member churches in their advocacy campaigns around ecological debt and climate change, mindful of the unity of God's creation and of the need for collaborative working between Southern and Northern nations. Specifically *supports* the activities of churches in countries that are suffering from climate change.

G. *Calls* for continued awareness-building and theological reflection among congregations and seminary students on a new cosmological vision of life, eco-justice and ecological debt through study and action, deeper ecumenical and inter-faith formation, and through the production and dissemination of relevant theological and biblical study materials.

H. *Urges* WCC member churches and church institutions to conduct ecological debt audits in partnership with civil society, including self-assessment of their own consumption patterns. Specifically, the WCC should establish a mechanism to provide for recompense of ecological debt incurred by its gatherings, and to collect positive examples of ecological debt recognition, prevention, mitigation, compensation, reparation and restitution in partnership with civil society groups and movements.

I. *Calls* for deepening dialogue on ecological debt and the building of alliances with ecumenical, religious, economic and political actors and between the churches in Southern and Northern countries.

J. *Stresses* the importance of accompanying ongoing struggles and strategically linking and supporting the efforts of peasant, women's, youth and indigenous peoples' movements through the World Social Forum and other avenues to design alternative compensation proposals, as well as to avoid amassing more ecological debt.

K. *Calls* upon WCC member churches through their advocacy work to encourage their governments to work for the recognition of the claims of ecological debt, including the cancellation of illegitimate financial debts.

L. *Calls* upon WCC member churches to deepen their campaigns on climate change by including climate debt and advocating for its repayment by applying the ecological debt framework.

M. *Calls* upon WCC member churches to advocate for corporate social accountability within international and national legal frameworks and to challenge corporations and international financial institutions to include environmental liabilities in their accounts and to take responsibility for the policies that have caused ecological destruction.

N. *Calls* upon WCC member churches to support community-based sustainable economic initiatives, such as producer cooperatives, community land trusts and bio-regional food distributions.

O. *Encourages* churches all over the world to continue praying for the whole of creation as we commemorate on 1 September this year the 20th anniversary of the encyclical of His All Holiness the Ecumenical Patriarch Dimitrios I, establishing the day of the protection of the environment, God's creation.

Prayer

The following prayer is offered as a resource to enable the churches' engagement with the issue articulated above:

Creator and creating God,

> *in the wonder of your world we experience your providential care for the planet and its people.*

We offer you our thanks and praise.

Creator and creating God,

> *in the exploitation of your world we recognize our human-centeredness and greed.*

We confess our sin before you.

We acknowledge our need for each other as part of your global family from North and South

And so we pray, "Forgive us our debts, as we forgive our debtors."

Accept our confession O God and offer us your forgiveness

empowering us to transform our lives as individuals, churches and nations,

proclaiming your love for the earth and its people,

enacting the principle of 'Jubilee' in our relationships with one another and the earth,

repaying our ecological debts in ways in ways which affirm your justice and shalom.

Appendix 3

Statement on Just Finance and the Economy of Life

And Jesus said to them, "Take care! Be on your guard against all kinds of greed; for one's life does not consist in the abundance of possessions." (Luke 12:15 NRSV)

1. The World Council of Churches (WCC) first articulated its concerns about finance and economics in 1984 when it issued a call for a new international order based on ethical principles and social justice. In 1998, the WCC assembly in Harare mandated a study on economic globalization together with member churches. WCC worked closely with the World Alliance of Reformed Churches, the Lutheran World Federation, Aprodev and other specialized ministries. Out of this, the Alternative Globalization Addressing People and earth (AGAPE) process, which was set up to further study the topics of poverty, wealth and ecology, was born. During the course of this process, several issues relating to various crises were identified: climate change and the food, social, and financial crises. In May 2009, the WCC convened a meeting of the Advisory Group on Economic Matters (AGEM) to (1) discern what is at stake in the current financial architecture, (2) propose a process that could lead to a new financial architecture and (3) outline the theological and ethical basis for such a new architecture.

2. Jesus warns that "You cannot serve both God and wealth" (Luke 16:13 NRSV). We, however, witness greed manifested dramatically in the financial and economic systems of our times. The current financial crisis presents an opportunity to re-examine our engagement and action. It is an opportunity for us to discern together how to devise a system that is not only sustainable but that is just and moral. Economics is a matter of faith and has an impact on human existence and all of creation.

3. The financial system of recent times has shaped the world more than ever before. However, by becoming the engine of virtual growth and wealth, it has enriched some people but has harmed many more, creating poverty, unemployment, hunger and death; widening the gap between rich and poor; marginalizing certain groups of people; eroding the whole meaning of human life; and destroying ecosystems. There is a growing and sobering awareness of our common vulnerability and of the limits of our current way of life. Today's global financial crisis, which originated in the richest parts of our world, points to the immorality of a system that glorifies money and has a dehumanizing effect by encouraging acquisitive individualism. The resulting greed-based culture impoverishes human life, erodes the moral and ecological fabric of human civilization, and intoxicates our psyche with materialism. The crisis we face is, at the same time, both systemic and moral. Those most affected are: women, who bear a disproportionate share of the burden; young people and children, as doubts are raised and their sense of security for the future is eroded; and those living in poverty, whose suffering deepens.

4. In an era of financial globalization, economic expansion has been increasingly driven by greed. This greed, a hallmark of the current financial system, causes and intensifies the sacrifice and suffering of impoverished human beings, while the wealthy classes multiply their riches. Finance is, at best, the lubricant of real economic activities. However, we note that money is not wealth; it has no inherent value outside the human mind. When it is turned into a series of fictitious instruments to create ever more financial wealth it is increasingly divorcing itself from the real economy, thereby creating only virtual or phantom wealth that does not produce anything to meet real human needs.

5. The abuse of global finance and trade by international businesses costs developing countries more than $160 billion a year in lost tax revenues – undermining desperately needed public expenditures. Developing countries are lending their reserves to industrial countries at very low interest rates and are borrowing back at higher rates. This results in a net transfer of resources to reserve currency

countries that exceed more than ten times the value of foreign assistance, according to the United Nations Development Programme (UNDP). This global financial crisis is proving the bankruptcy of the neoliberal doctrine, as promoted by the International Financial Institutions through the "Washington Consensus." The leaders of the rich countries that had promoted the consensus so emphatically, declared it "over" at the G20 meeting in April 2009. And yet much of the G20's agenda reflects misguided efforts to restore the same system of overexploitation of resources and unlimited growth. Furthermore, resources are channeled through the militarization of some societies, due to a perverse understanding of human security through military power.

6. Unfortunately, churches have also been complicit in this system, relying on popular models of finance and economics that prioritize generating money over the progress and well-being of humanity. These models are largely oblivious to the social and ecological costs of financial and economic decisions, and often lack moral direction. The challenge for churches today is to not retreat from their prophetic role. They are also challenged by their complicity with this speculative financial system and its embedded greed.

7. There are two structural elements of the current paradigm which must be changed. First, the economic motive of surplus value, unlimited growth and the irresponsible consumption of goods and natural resources contradict biblical values and make it impossible for societies to practice cooperation, compassion and love. Second, the system that privatizes productive goods and resources, disconnecting them from people's work and needs and denying others access to and use of them is a structural obstacle to an economy of cooperation, sharing, love and dynamic harmony with nature. Alternative morality for economic activity is service/koinonia (fellowship) to human needs; human/social self-development; and people's well-being and happiness. An alternative to the current property system is connected to need, use and work invested in the production and distribution process. In order to achieve this goal, the existing organizing principles of production and claims settlements (i.e., distribution) must change. This also warrants a

situation where an ethical, just and democratic global financial architecture emerges and is grounded on a framework of common values: honesty, social justice, human dignity, mutual accountability and ecological sustainability. It should also account for social and ecological risks in financial and economic calculation; reconnect finance to the real economy; and set clear limits to, as well as penalize, excessive and irresponsible actions based on greed.

8. It is in this context that the central committee of the WCC acknowledges that a new ethos and culture which reflects the values of solidarity, common good and inclusion must, at this time of crisis, emerge to replace the anti-values of greed, individualism and exclusion. New indicators of progress, other than Gross Domestic Product, such as the Human Development Index, the Gross National Happiness (GNH) index and ecological footprints and other corresponding systems of accounting need to be evolved. For example, a GNH index that reflects the following values: 1) Quality and pattern of life; 2) Good governance (true democracy); 3) Education; 4) Health; 5) Ecological resilience; 6) Cultural diversity; 7) Community vitality; 8) Balanced use of time; 9) Psychological and spiritual well-being.

9. The central committee of the WCC also emphasizes the need for a new paradigm of economic development and a re-conceptualization of wealth to include relationships, care and compassion, solidarity and love, aesthetics and the ethics of life, participation and celebration, cultural diversity and community vitality. This will involve responsible growth that recognizes human responsibility for creation and for future generations – an economy glorifying life.

In view of the need to support international organizations that are democratic, to represent all member nations of the United Nations (UN) and to affirm common values, the central committee of the WCC, meeting in Geneva, Switzerland, 26 August - 2 September 2009, calls upon governments to take the following necessary actions:

A. *Adopt* new and more balanced indicators, such as the Gross National Happiness (GNH) index, to monitor global socio-environmental / ecological-economic progress.

B. *Ensure* that resources are not diverted from basic education, public health, and poor countries.

C. *Uphold* their commitments to and assistance for meeting the Millennium Development Goals (MDGs), particularly the goal number 8 on cooperation world-wide.

D. *Implement* gender-just social protection programmes as an important part of national fiscal stimulus packages in response to the current financial crisis.

E. *Emphasize* the participation of people and civil society organizations in policy-making processes, including the promotion of decentralized governance structures and participatory democracy.

F. *Treat* finance also as a public service by making loans available to small and medium enterprises, farmers and particularly poor people through, for example, micro-financing in support of not-for-profit enterprises and the social economy.

G. *Support* regional initiatives that decentralize finance and empower people in the global South to exercise control over their own development through such proposed bodies as the Bank of the South, the Asian Monetary Fund and the Bank of ALBA.

H. *Revise* taxation systems, recognizing that tax revenues are ultimately the only sustainable source of development finances, by establishing an international accounting standard requiring country-by-country reporting of transnational companies' economic activities and taxes paid and by forging a multilateral agreement to set a mandatory requirement for the automatic exchange of tax information between all jurisdictions to prevent tax avoidance.

I. *Explore* the possibility of establishing a new global reserve system based on a supranational global reserve currency and regional and local currencies.

J. *Achieve* stronger democratic oversight of international financial institutions by making them subject to a UN Global Economic Council with the same status as the UN Security Council.

K. *Explore* the possibility of setting up a new international credit agency with greater democratic governance than currently exists under the Breton Woods institutions.

L. *Set up* an international bankruptcy court with the authority to cancel odious and other kinds of illegitimate debts and to arbitrate other debt issues.

M. *Regulate* and *reform* the credit agency industry into proper independent supervision institution(s), based on more transparency about ratings and strict regulation on the management of conflict of interest.

N. *Use* innovative sources of finance, including carbon and financial transaction taxes, to pay for global public goods and poverty eradication.

Prayer

The following prayer is offered as a resource to enable the churches' engagement with the issue articulated above:

O God who is one in Trinity, in you we find the perfect relationship of love and justice.

We confess:

that too often our relationships have been characterized by greed and self-interest,

that we have sought wealth and security for ourselves with little thought for your creation,

that our desire for more has meant that others have less,

that we have displayed the Pharisees arrogance and not the widow's sincerity in our giving.

Inspire us with a vision of your oikumene, characterized by love and compassion:

where all have enough to eat,

where work is justly rewarded,

where concern for the least is our most pressing demand,

where life is celebrated and you, the giver of life, is praised.

Appendix 4

Statement on the Current Financial and Economic Crisis with a Focus on Greece

At the present time your plenty will supply what they need, so that in turn their plenty will supply what you need. The goal is equality…
(2.Cor. 8. 14, NIV)

1. We live in an interconnected and interdependent world which is experiencing more than ever before, a severe financial crisis. This crisis is due to various reasons such as unjust economic and financial policies; structural weaknesses of political, economic and financial institutions; and a lack of ethical values in a world that is increasingly dominated by the greed of the powerful seeking short-term advantages and maximum profit and denial of the need of the powerless. The events since 2008 have caused severe strains in the global economy, and have strained public finances even as the millions of people who lost their jobs, pensions and homes in the aftermath continue to clamour for social protection.

2. Europe has been at the centre of the most recent economic problems, and the immediate challenge for the Euro zone is the financial crisis in Greece. In addition, Italy and Spain, the third- and fourth-largest economies in the euro zone respectively, represent another major problem, with investors pushing the interest rates on their bonds to unsustainable levels. The fear is that financial instability in the Euro zone will provoke another global panic similar to – and potentially graver than – the one in 2008 with adverse consequences for socio-economically weak nations and peoples.

3. Triggered by the 2008 global financial fallout, Greece's debt problem arose partly from government mismanagement, but blame also attaches to irresponsible lenders that offered easy loans and stimulated housing bubbles, regulators that failed to regulate, and political leaders who were blind to the challenges of establishing a single European currency system among diverse economies. In Greece, the harsh austerity packages aimed

at stabilizing markets and satisfying international creditors have created a new "underclass" of the unemployed, homeless and hungry. Since 2010, taxes have been raised especially indirect taxes of up to more than 20 per cent on food; pensions and state salaries slashed across the board, and retrenchment of public workers in tens of thousands. An alarming unemployment rate, more than 20 percent in general and more than 50 percent among the youth, has caused deteriorating living standards leading to frustration, anger and violence, especially against immigrants. Public nursing programmes for the elderly have been shut down. Small businesses are being forced to close or struggle to survive. Women's unpaid labour is substituting for cutbacks in social programmes. Suicides have spiraled in the last couple of years.

4. The ministries of many church congregations are being directly challenged and affected by these changes. As an example, many churches' feeding and shelter programmes are struggling to keep up with the growing numbers of people availing their services; and the spiritual and pastoral needs of those experiencing these challenges in their families are increasingly profound.

5. Despite many severe measures, Greece's debt has not been brought under control. Tax evasion is a significant issue in certain sectors of Greek society. Austerity is resulting in a vicious cycle of economic decline, hampering recovery by dampening domestic demand and eroding national tax revenues, and therefore making it even more challenging for the country to settle its debt. There is no justice when those who had little part in generating the crisis pay the highest price for it. It is immoral to demand austerity and debt repayment at human and social cost which falls unfairly on the weaker members of society. Moreover, there is a need for a healthy approach to creativity, personal and corporate financial responsibility for the sake of the common good, productivity and small business in order to create the optimal conditions for the exercise of generosity, compassion and justice.

6. The World Council of Churches (WCC) has been closely observing the global financial situation since the unraveling of financial markets in 2008 and has issued letters addressed to the United Nations General Assembly and the Group 20 as well as statements calling on

governments to go beyond short-term measures and to address the roots of the financial and economic crisis.

7. We believe that reforming the international financial and monetary systems in the context of global public authority is an urgent priority. We need to be engaged in a process of searching for a viable model of sustainable development and associated financial systems.

8. In this painful financial crisis, the church is being called upon to defend the dignity of all people, as made in the image of God. The crisis is spiritual and moral, as well as economic. The Christian values of justice and love have a renewed importance in Europe today. The excessive differences between the wealthy and the poor, and the growing levels of unemployment, especially among young people, which have developed in recent decades are immoral, and will not form the basis for a healthy society. The church is bound to believe that current events embody a message from God, and will give us an opportunity for discernment to shape our visions for a better future of equality and justice to all God's people.

The Central Committee of the World Council of Churches, meeting in Kolympari, Crete, Greece, from 28 August to 5 September 2012, therefore:

A. *Affirms* its solidarity with the people of Greece, and others who are particularly suffering from the current crisis

B. *Reiterates* our call for economic policies which do not encourage irresponsible debt, national or private, and which spread the benefits of wealth more fairly to all citizens, especially the weak and marginalized, including young people ;

C. *Urges* the prevention of the recurrence of crises in the future, by the regulation and restructuring of the banking industry, and continuation of search for deep-seated transformations in the current international financial regime, as outlined in the WCC Central Committee statement on "Just Finance and an Economy of Life," issued in September 2009;

D. *Supports* the principle of a financial transaction tax (FTT) as a sensible tool that will enable governments to meet their obligation to protect and fulfill the economic, social and cultural rights of their people. The FTT would not only help to curb speculation, but would also ease sovereign debt loads, transfer the burden from ordinary people to the private sector which set off the crisis in the first place, and considerably expand government fiscal space for spending on urgently needed social protection policies;

E. *Calls* up on Churches in this time of crisis to address these issues with a particular focus of talking to power on the one hand and seeking ways of supporting those who are now marginalized by the current financial policies on the other, and commend them for their ongoing attention to the spiritual and pastoral needs of those, including youth, whose lives will be most directly affected by these troubling economic challenges;

F. *Urges* the Churches in Europe to stand together and to advocate for common European solutions to the financial and social crisis that help to deepen the project of European Unity as a project of just peace on the continent;

G *Invites* churches and faith based organizations to continue to mobilize and to support one another for the immediate relief and assistance of the weakest members of our society.

Appendix 5

The São Paulo Statement
International Financial Transformation for the Economy of Life

Global Ecumenical Conference on a New International Financial and Economic Architecture, 29 September - 5 October 2012, Guarulhos, State of São Paulo, Brazil

From its inception, the ecumenical movement has critically engaged with issues of economic and social justice. In particular, the current global economic crisis, which also affected rich economies in 2008 and has thrown millions of people across the globe into poverty, has grasped our attention. In response, the World Council of Churches (WCC) addressed the United Nations and the Group of 20 (G20), calling on governments to tackle systemic greed and inequality. In 2009, the WCC issued a "Statement on Just Finance and the Economy of Life" calling for an ethical, just and democratic international financial regime "grounded on a framework of common values: honesty, social justice, human dignity, mutual accountability and ecological sustainability." [1] *In 2010, as part of the commitment to live out the Accra Confession*[2]*, the Uniting General Council of the World Communion of Reformed Churches (WCRC) called upon its members, in partnership with the WCC and other ecumenical bodies, to prepare an international ecumenical conference to propose a financial and economic architecture that:*

is based on the principles of economic, social and climate justice;

serves the real economy;

accounts for social and environmental tasks; and

sets clear limits to greed.

---

[1] See http://www.oikoumene.org/en/resources/documents/central-committee/geneva-2009/reports-and-documents/report-on-public-issues/statement-on-just-finance-and-the-economy-of-life.html.

[2] See http://www.wcrc.ch/sites/default/files/Accra%20Conf%20ENG_0.pdf .

Sharing a deep commitment to promoting justice in the economy and the Earth and recognizing the need to work together to have a meaningful impact, the WCC, WCRC and Council for World Mission (CWM) convened the Global Ecumenical Conference on a New International Financial and Economic Architecture to engage the proponents of diverse proposals and solutions, set criteria and a framework and develop a plan of action toward constructing just, caring and sustainable global financial and economic structures.

We - economists, church leaders, activists, politicians and theologians - gathered in the State of São Paulo, Brazil, between 29 September and 5 October 2012 to envision together an alternative global financial and economic architecture. The gathering was a response to and continuation of the decades of work around issues of economic, social and ecological justice with which the WCRC, WCC, CWM, and the Lutheran World Federation (LWF) have been involved. The visions and the criteria for a new financial and economic architecture and the alternatives that are expressed in this document therefore build on the Accra Confession of the WCRC, the "Statement on Just Finance and the Economy of Life" and the AGAPE Call to Action "Economy of Life: Justice and Peace for All" [3] of the WCC as well as the theological statement on "Mission in the Context of Empire" of CWM[4].

The 2008 global financial and economic crash increased poverty and unemployment among millions in the global North and worsened and deepened poverty, hunger and malnutrition among even larger numbers in the global South, already experiencing decades of poverty and deprivation caused by injustices in international financial and economic relations. A system of speculation, competition and inadequate regulation has failed to serve the people and instead has denied a decent standard of life to the majority of the world's population. The situation is urgent.

---

[3] See http://www.oikoumene.org/en/resources/documents/wcc-programmes/public-witness-addressing-power-affirming-peace/poverty-wealth-and-ecology/neoliberal-paradigm/agape-call-for-action-2012.html.
[4] See http://www.cwmission.org/wp-content/uploads/2012/12/CWM-Theology-Statement-2010-final.pdf.

Critical theological reflection on the material and collective bases of life has been intrinsic to the call to be faithful disciples of Christ and has expressed itself through theological contemplative praxis that has sought transformative liberation from unjust socio-political, cultural and economic structures, thereby promoting the fullness of life for all creation.

Modernity has, however, brought with it an economic model based on profit and self-interest disconnected from faith and ethics. This has led to the ideological justification of colonialism, the despair of poverty and inequality, and the violence of economic and ecological devastation as well as the reluctance of some churches to discern the signs of the times and to engage with the realities of a dehumanizing dominant world order that continually discriminates and oppresses those with whom God sides: the poor, the downtrodden, the disadvantaged and the oppressed.

The immersion visits in São Paulo exposed the narratives of the homeless, the single mother, the widow, the orphan, the addict, as representing just some of the disenfranchised. This was a visible encounter with those whom society has left on the periphery. Patriarchal perceptions, racist subjugative ideologies, anthropocentric domination and discriminative comprehensions of the human hierarchical order induced by the sin of neoliberalism, supported by heretical theology which justifies it, and legitimized by the idolatry of imperial globalization have perverted relationships between God, human beings and the Earth.

The God of the oppressed calls us into an alternative imagination which has to emerge from the margins, from those who have been left out of socio-political and economic decision making but are the first to suffer its consequences.

We therefore seek a transformative theological praxis that not only delegitimizes, displaces and dismantles the present social and economic order but also envisions alternatives that emerge from the margins. There is thus a requirement for an active radicalizing of our theological discourse that will no longer allow too much power being placed into capitalist ideologies that have resulted in an inability to think beyond existing financial and economic structures.

This alternative imagination has to be derived from our spiritual and theological convictions, employing liberative theologies that respond to concrete systematic struggles, inclusive of feminist, womanist, mujerista, eco-feminist, Latin American liberation, black, ecological, post-colonial, grass-root, minority and public theology, and indigenous spiritualities. The list of hermeneutical lenses of suspicion and retrieval required to bring about transformative change continues to be as extensive as the list of those who have been downtrodden and persecuted by the dominant economic world order.

We lament the manner in which economic and financial legislation and controls are biased in favour of the wealthy. We therefore affirm the God of justice for all those who are oppressed (Ps. 103:6). We call for a system of just legislation and controls that facilitate the redistribution of wealth and power for all of God's creation.

Therefore, we reject Empire[5] and our complicity with all systems of death, including militarism, and affirm movements of social concern and other radical traditions that are a rejection of Empire and seek to build life in community outside the logic of hierarchy and discrimination.

We reject political and military offences perpetrated in the name of neoliberalism that threaten human security and result in massive violations of human rights.

Therefore, we reject the explosion of monetization and the commodification of all of life and affirm a theology of grace which resists the neoliberal urge to reduce all of life to an exchange value (Rom. 3:24). Means have become ends; instruments have become a means for the coercion of facts.

We reject an economy that is driven by debt and financialization in favour of an economy of for-giveness, caring and justice and declare that debt and speculation have reached their limits. We affirm the words of

---

[5] "In using the term 'empire' we mean the coming together of economic, cultural, political and military power that constitutes a system of domination led by powerful nations to protect and defend their own interests" (Accra Confession).

the Lord's Prayer in which we pray to have our own debt forgiven in the same manner as we forgive the debts of others (Matt. 6:12).

Therefore, we reject the ideology of consumerism and affirm an economy of Manna, which provides sufficiently for all and negates the idea of greed (Ex. 16).

We reject increasing individualistic consumerism by affirming and celebrating the diversity and interconnectedness of life. We further affirm that wholeness of life can be achieved only through the interdependent relationships with the whole of the created order. The idea of a Triune God acts as a challenge to individualism, discrimination and exclusivity; it is a doctrine that calls us into a life of equality in community and requires an active response that affects the whole of humanity.

Based on the moral principle of the diversity of the cosmos, we therefore exclude notions of exclusivity by promoting and affirming the need for interfaith dialogue. This requires a praxis of connectivity enabling a wider dissemination of spiritual resources gathered from faith communities, inclusive of the insistence of the Qur'an on the rejection of interest, the valorization of moral banking and a concentration on the real economy, as well as the Islamic injunction on limits to consumption that is expressed through the idea of Ramadan and fasting, and that resonates with the way in which many Christians around the world practice Lent.

We reject an economy of over-consumption and greed, recognizing how neoliberal capitalism conditions us psychologically to desire more and more, and affirm instead Christian and Buddhist concepts of an economy of sufficiency that promotes restraint (Luke 12:13-21), highlighting, for example, the Sabbath economy of rest for people and creation, and the Jubilee economy of redistribution of wealth.

We reject the economic abstraction of *Homo Oeconomicus*, which constructs the human person as being essentially insatiable and selfish, and affirm that the Christian perception of the human person is embedded in community relationships of Ubuntu[6], Sansaeng[7], Sumak Kawsay[8], conviviality and mutuality. Contrary to the logic of neoliberals, as believers we are called to think not only of our own interests but also of the interests of others (Phil. 2:4).

We acknowledge our role in the destruction of the Earth's resources and the impact this has had on the vulnerable nations in the South. We continue to seek forgiveness through practical actions and solutions that militate against ecological destruction.

We affirm ourselves as prophetic witnesses, as we have seen the injustices and structural violence of our age and those of a history of domination. We have discerned the signs of the times in the light of our calling as disciples of Jesus. Therefore we seek to overcome capitalism, its nature and its logic and to establish a system of global solidarity. We search for alternatives, for just, caring, participatory and sustainable economies such as a solidarity economy and gift economy.

We affirm that the only choice that Jesus offers us is between God and Mammon (Matt. 6:24), as those who desire to be faithful followers; we have no choice but to do justice, love mercy and walk humbly (Micah 6:8).

Therefore, we present the following criteria and framework:

Criteria and Framework

We are called to find a new and just international financial architecture oriented toward satisfying the needs of people and the realization of all economic, social and cultural rights and human dignity. Such architecture must be focused on reducing the intolerable chasm between the rich and the poor and on preventing ecological destruction. This requires a system which does not serve greed but which embraces

---

[6] "Ubuntu is an African concept of personhood in which the identity of the self is understood to be formed interdependently through community" (Michael Battle, *Ubuntu: I in You and You in Me*, Seabury Books, New York, 2009, p.1f).
[7] Sansaeng is the Korean concept of "life together" (*International Review of Mission*, April 2012, p.15).
[8] Sumak Kawsay in Quichua is the notion of "good living" or "good life."

alternative economies that foster a spirituality of enough and a lifestyle of simplicity, solidarity, social inclusion and justice.

Overcoming greed: The drive to consume is a culture of greed that destroys all of God's creation. The economic activity of the last five centuries has caused massive ecological destruction. Over the years, big businesses, governments and multinational corporations have been reckless through policies and practices of unlimited growth which have led to pollution, destruction of forests, overproduction and the alienation of the poor and of farm workers from the land. Natural resources are limited, and the human ecological footprint already surpasses the Earth's bio-capacity. Individual self-interest and long-term social welfare are not necessarily compatible, and market mechanisms do not lead to an optimal social distribution. Therefore, political regulation is required to optimize sustainable social welfare.

Social inclusion: There is a distorted definition of anthropology in neoliberalism in which human beings are defined by financial and economic value and not by their intrinsic dignity as persons created in the image of God. This anthropology has nested in humanity, colonizing our mind and our dreams. This definition leads to racism, sexism and other forms of categorization, exclusion and oppressive behaviour. This is a sin against God, humanity and all creation.

Gender justice and ecological justice: We need an economy that recognizes the link between gender justice and ecological justice. The degradation of the land and Earth has dire consequences for the lives of the marginalized, especially the poor, women and children in poor countries. Land is tied closely to women both physically and symbolically. Physically, women till the land and walk miles for water for their families. Symbolically, the sufferings of the land are likened to the pains and groans of a woman at childbirth (Rom. 8:22). To put it differently, the "economy of care" for the Earth cannot be separated from the issue of justice for all of God's creation.

Hope: We are committed to affirming existing alternatives to neoliberal capitalism. Persons living in poverty and deprivation as a result of neoliberal financial systems have demonstrated that alternative life-giving economies are alive, impacting millions of indigenous and grassroots people. It is to these initiatives that we must turn for criteria that truly speak to an alternative. Throughout the world, people's movements resist the temptation to surrender to a death-dealing economic system. At the same time, many poor and marginalized people survive through a variety of systems which, even though not recognized by big business, governments and mainstream economies, nevertheless keep them alive and nurture hope.

Spirituality and economy: There is a need to democratize and demystify economic knowledge and to free public imagination to promote social and personal wellbeing on a foundation of economic justice. There are Biblical, Qur'anic and indigenous narratives that point us to economic life-giving systems where there is equitable sharing, communion with creation, abundance for all and affirmation for the fruits of our labour as offerings for the common good. Biblical motifs such as Jubilee, Shalom, Eucharist, *oikos*, and *Koinonia* remind us of God in community with God's creation as well as the covenantal relationship into which God invites us (cf. Ex. 16:16-21).

The dire crises with which we are confronted imply that our long-term vision has to be accompanied by short- and medium-term goals; therefore, we recommend the following:

An Ecumenical Plan of Action and Landmarks of a New International Financial and Economic Architecture

The world economy and the international financial system have become globalized but democratic governments have not followed in any appropriate way. The key democratic problem is the lack of sovereignty over our lives, the planet and the future. Markets rule. As a result, we see a patchwork of governing systems with overlapping and often competing competencies. Some of these suffer from a serious deficit of justice and lack of democratic credentials. The G20 constitutes a case in point, where a group of self-appointed world leaders discuss and decide issues that affect far more people than those who live in their own countries. By the same token, the International Financial Institutions

(IFIs) are not based on a democratic system. Rather, their decision-making structures reflect the relative economic and financial power of nation states.

In order to address these inequalities, nothing less than a drastic overhaul of the governance of the world economy and the international financial system is needed. The main objective is to ensure that financial markets and the economy are brought under the primacy of democratic decision-making structures and that they function as good servants rather than bad masters in political and economic life. Economics has to be embedded in social, ecological and political life rather than the other way around.

This plan identifies policies to address the fundamental issues mentioned above, distinguishing between immediate and medium-term actions, and longer-term structural changes of the global economic and financial architecture. Furthermore, we suggest a number of strategies for the churches and recall signs of hope.

Immediate and Medium-Term Actions

Alternative indicators of economic wellbeing: Governments and international institutions should replace growth in Gross Domestic Product (GDP) as the primary indicator of economic progress by other indicators, including growth of decent work, qualitative as well as quantitative indicators of health and education, and measures of environmental sustainability.

Regulating the financial sector: A number of measures are necessary to regulate and transform the financial sector:

- A comprehensive regulation of the entire financial sector, including the lightly regulated shadow banking sector (which in the US and Europe is larger than the banking sector) is required.

- There is a need to ensure that banks have adequate capital to absorb losses. Regulations on permitted leverage and minimum liquidity must be rigorous; likewise, counter-cyclical prudential regulation can assist in macroeconomic management.

- Basic banking activities of deposit taking and lending to enterprises and households should be tightly regulated and separated from more risky activity (as in the United States in the 1930s with the Glass-Steagall Act).

- Banks that are "too big to fail" should be broken up.

- Speculative activity should be restricted so that the counterpart to real-economy hedging needs is met without overwhelming enterprise on a "sea of speculation." Regulators should set "position limits" on commodity traders in all globally relevant markets, especially those of foodstuffs, to limit unnecessary price volatility. Regulators should also require that market participants are capable of accepting delivery of the actual commodities. Further Credit Default Swaps, which have played a harmful role in the recent financial crisis, should be banned.

- Public policy should be directed to the reform of bankers' remuneration systems, to link them to long term social and environmental performance rather than short-term results. For example, bonuses could be set at a maximum of 100 percent of fixed remuneration (as demanded by the European Parliament). Commission should be forbidden when selling financial products to retail investors.

Financial Transaction Tax: A global Financial Transactions Tax on trades in equities, bonds, currencies, and derivatives should be established immediately. Likewise, a democratically representative agency to receive and allocate the proceeds for global public goods, including the eradication of poverty and disease, and the costs of climate change mitigation and adaptation incurred by low-income countries, must also be set in place.

Ensuring access by poor and marginalized sectors to basic financial services: In line with the principle that finance should be a valuable public service, financial services such as savings accounts and credit must be made available on acceptable terms to small and medium enterprises, people in poverty, women and farmers. The setting up of credit unions should be encouraged to provide productive loans to sectors of society that are not deemed creditworthy by the mainstream banking industry, often on account of poverty, class, gender and race.

Investment and sustainable development policies: Governments have a role in supporting long-term, socially useful investment through strong investment programmes for renewable energy, sustainable agriculture and energy efficiency. Governments should also set binding emission caps for greenhouse gases and binding product norms (such as moving caps for energy efficiency for buildings), support sustainable investment banks and social finance institutions, and make green technology available to the South.

Progressive taxation: Capital gains must be taxed in the same way as other income. Likewise income taxes should be made much more progressive, especially for the highest income earners. Revenues from wealth taxes and estate taxes should be used for public purposes.

Gender-just fiscal stimulus and social protection: Public investment and spending on small-scale agriculture, renewable energy, infrastructure, health and education sectors, and gender-just social protection programmes must be safeguarded and expanded even during periods of painful austerity measures in debt-burdened nations. Austerity often falls heavily on the most vulnerable sectors of society and results in a vicious circle of economic decline, hampering recovery by dampening domestic demand and eroding national tax revenues.

Addressing tax evasion and avoidance: A multinational framework for the compulsory exchange of tax information on individual and corporate accounts that will effectively end the use of tax havens must be established. Transnational corporations should be required to report sales, profits and taxes paid on a country by country basis in their audited financial reports.

Ecological taxation: Ecologically destructive industries and activities must be heavily taxed or prohibited. Fossil fuel extraction and carbon emissions should be taxed while at the same time rebating some of the proceeds to low-income households and using other revenues for investments in energy efficiency, conservation and renewable energy to assist in the transition to a low-carbon economy.

Regulating financial flows for sustainability: Governments should be encouraged to manage capital flows so that surges or flows in or out of a country do not destabilize the economy, including through instruments such as capital controls. Capital controls could curb the entry of volatile short-term flows as well as prevent capital flight from already beleaguered economies.

Sovereign debt restructuring mechanism: A comprehensive, fair and transparent international debt restructuring mechanism to address sovereign insolvency on a timely basis should be established. Such a mechanism must be empowered to audit sovereign debts and cancel those debts found to be odious because they were contracted by despotic regimes without public consent for use against the population, or are illegitimate due to other factors such as usurious interest charges, fraud, and repayment obligations that would cause unacceptable privation.

Information and communication architecture: Information and communication structures that support alternative financial and economic structures must be promoted.

Structural Changes

United Nations Economic Social Ecological Security Council: For all its deficiencies, the United Nations remains the most representative and inclusive forum for global cooperation and policy setting. Conceptually, it serves as a model on which to build a more effective and representative international financial and economic architecture. However, it is not adequately forging consensus on many issues at this time.

A potential instrument for enhanced, effective and coherent global governance could be the establishment of a UN Economic, Social and Ecological Security Council (UNESESC). Civil society and churches have repeatedly called for such a body where pressing economic, social and ecological issues would be brought together to be discussed and acted upon in a coherent way. The report of the Stiglitz Commission, published in 2009, echoed this demand[9]. As proposed by the Stiglitz Commission, the task of the UNESESC would be to assess developments and provide leadership in addressing economic issues that require global action while taking into account social and ecological factors. It should represent all regions of the world at the highest possible level and ensure the participation of the various global institutions (such as the IFIs, International Labour Organization, United Nations Conference on Trade and Development, United Nations Women, World Health Organization, United Nations Development Programme, United Nations Educational Scientific and Cultural Organization, International Telecommunication Union, etc.), and cooperate closely with civil society to promulgate measures for the protection of the economic, social and ecological rights of nations and communities.

A church-led initiative should bring the interested stakeholders together to develop the proposal further in order to overcome differences that impede reaching the consensus needed for implementation. In addition and as an interim measure, an informal intergovernmental forum at the UN could bring representatives of governments, multilateral institutions, private sector and civil society organizations together in order to build consensus on financial policy and governance reforms that serve society.

A new International Monetary Organization to replace the International Monetary Fund: A new International Monetary Organization (IMO) needs to be created and should be guided by universal principles of economic, social and ecological justice. The IMO would have oversight over monetary policies and transactions and would deploy funds without structural adjustment conditions to establish a globally effective, stable, fair and socially responsible global financial and economic architecture, bringing democratic accountability to financial markets. Its actions should not be dominated by policies of interest groups and its policies should be equitable and responsive to the social consequences of financial activities at financial sector and national levels.

---

[9] "Report of the Commission of Experts of the President of the United Nations General Assembly on Reforms of the International Monetary and Financial System, September 21, 2009," New York, United Nations. See http://www.un.org/ga/econcrisissummit/docs/FinalReport_CoE.pdf.

The proposed IMO should direct its policies toward economies in the service of life. Its policies should cover areas such as capital flows, control of capital flight, taxes on capital flows, and, where and when appropriate, (re)establishing fixed currency exchange rates. Such policies would enable countries to regain autonomy of fiscal and monetary policy. Furthermore, the IMO should deal with limiting excessive speculation, fair burden sharing of private creditors in dealing with the impacts of financial crises, increased cooperation in financial market controls, guidelines for risk management, closing tax havens, etc.

An alternative international reserve currency: There is a need to design a new multicurrency reserve asset, similar to Special Drawing Rights, to create liquidity so that the "seigniorage" currently enjoyed by those countries whose currencies are now used as reserves instead accrues to the international community. At present, the main commonly used international reserve currency is the US dollar. Almost everywhere in the world, the US dollar is accepted and convertible. This creates enormous advantages for the US economy as, contrary to other countries, the United States can pay for some of its imports with dollars instead of with exports, as long as the world considers the dollar a safe reserve currency. No other country in the world would survive with a level of current account deficits as high and as persistent as those of the US. This "seigniorage" is an "exorbitant privilege" which accrues to the US. It is a significant unjust feature of the present international financial system, coupled with the fact that there are often undesirable consequences for the world's economies, such as excessive capital flows, resulting from the monetary policies that the United States takes for purely domestic reasons.

In order to make the world less dependent on US deficits (or gold reserves, for that matter) and in order to create global liquidity in a more rational way, the International Monetary Fund created in the 1960s a multilateral reserve asset called Special Drawing Rights. SDRs can be created as the objective need arises, for example as an instrument for anti-cyclical policies (as in 2009), and as an alternative reserve asset which could eventually replace the US dollar and a few other reserve currencies. Besides SDRs, other proposals have been made such as International Currency Certificates. The common aim of these proposals is to search for ways and means to arrive at a system for the creation of liquidity based on global need in order to serve the real economy.

Strategies and Actions for Churches

To move forward the agenda outlined above, a global ecumenically instituted commission should be formed immediately to carry forward the valuable work of the Stiglitz Commission, linking with other faith communities, civil society organizations, interested governments, institutions and other relevant stakeholders to develop a concrete proposal for the governance of a new world economic and financial architecture.

Further, the following actions are recommended:

- The WCC, WCRC, CWM and LWF should, together with other partners, develop a coherent strategy of advocacy for a new economic and financial architecture. Effective communication strategies are key for successful advocacy initiatives.

- Churches should substantially increase the number of staff working on building dialogue on economic and financial developments with decision makers in the fields of politics, the private sector, professional associations, standard setting institutions, research organizations and civil society organizations.

- An ecumenical school of Governance, Economics and Management (GEM) should be established to develop economic competencies and empowerment within the ecumenical movement. In addition, educational materials should be developed to enhance the economic and financial literacy of church members.

- Churches should affirm a commitment to communication rights to advance the empowerment of communities in developing alternatives to the current financial and economic structures.

- The ecumenical movement should accompany alternative social movements from below that protest against the injustices of the present system and strive to develop alternatives (e.g. the World Social Forum and, more recently, the "Occupy" movement).

- As a matter of accountability, churches should be asked to report on how they have followed up on recommendations on ethical investments. Such responses could be used to strengthen ecumenical cooperation in this area.

Signs of Hope

The agenda for transformation is vast, and it is easy to be overwhelmed by all that is required to implement it. Yet numerous alternatives have already been established by people all over the world and that serve as signposts of change:

- Organizations and people are making a distinction between material wealth and wellbeing and are advocating that limits be observed for those who are already well off. In this context, efforts are being made to develop new indicators, such as the Human Development Index, in addition to or instead of GDP. The main world religions are well placed to contribute to these initiatives as they are inspired by ideas such as "life in all its fullness."

- Initiatives are being taken to promote education and financial literacy and to coach those who need accompaniment in dealing with money and finance.

- The various forms of provision of care services (unpaid, paid public and paid private) are key indicators that are as important as financial balances and production of tradable goods and services. Policies ought to be based on these indicators to enhance recognition of the vital role that women play in economic life, and should further promote the role of such indicators.

- Other concrete examples are Local Exchange Trading Systems (LETS) which create their own local economies, the setting up of credit unions, the use of ethical investments, and organizations such as Oikocredit and ECLOF, which are micro finance organizations established by the WCC and partners.

- Latin American integration and independence is advancing though political organizations such as the Bolivarian Alternative for the Americas (ALBA), Union of South American Nations (UNASUR) and Community of Latin American and Caribbean States (CELAC), and economic cooperation through MERCOSUR (a regional trade agreement), SUCRE (a regional currency), Banco del Sur and the proposed Fondo del Sur.

In view of the gross injustices that accompany neoliberal policies and structures, nothing less than a *metanoia* of the international economic and financial system is required. For that we need a people's movement which, like the earlier civil rights, anti-apartheid and Jubilee movements, rejects a world that is unfair, unequal and unjust, and one that is run for the benefit of the "1 percent."

Ultimately, changes will need to go beyond technical and structural requirements. What the world needs is a change of heart so that financial and economic systems do not have individual gain as their compass but justice, peace and the protection of God's creation.

Abbreviations

AGAPE – Alternative Globalization Addressing People and Earth

AGEM – Advisory Group on Economic Matters

CDR – Common, but Differentiated Responsibilities

CEC – Conference of European Churches

CONAVIGUA – Coordinadora Nacional de Viudas de Guatemala (National Coordinator of the Widows of Guatemala)

COP 15 – Conference of Parties

CLAI – Latin American Council of Churches

CSW – Commission on the Status of Women

CWM – Council for World Mission

ECADD – Europe-China-Africa Development Dialogue

EPA – Economic Partnerships in Africa

FECCIWA – Fellowship of Councils of Churches in West Africa

FOCCISA – Fellowship of Christian Churches in Southern Africa

GDP – Gross Domestic Product

GDR – Greenhouse Development Rights

GHG – Greenhouse Gas

ICCO – Inter-church Organization for Development Cooperation

IFI – International Financial Institutions

ILO – International Labour Organization

IEPC – International Ecumenical Peace Convocation

IMO – International Monetary Organization

LWF – Lutheran World Federation

MDG – Millennium Development Goals

POSCO – formerly Pohang Iron and Steel Company

PWE – Poverty, Wealth and Ecology

UNESESC – UN Economic, Social and Ecological Security Council

UNFCCC – United Nations Framework Convention on Climate Change

WARC – World Alliance of Reformed Churches

WB – World Bank

WSCF – World Student Christian Federation

WCC – World Council of Churches

WCRC – World Communion of Reformed Churches

WSF – World Social Forum